Kinetics and Mechanisms of Reactions

D1305008

Methuen Studies in Science

GENERAL EDITOR   J. M. Gregory M.A., D.Phil., Winchester College

CONSULTANT EDITORS   B. E. Dawson B.Sc., Ph.D., King's College, London

                        R. Gliddon B.Sc., Ph.D., Clifton College, Bristol

This series provides concise, introductory surveys of important topics in the physical, chemical and biological sciences. It is designed to help students in preparing for entry to unversity or college or for more advanced studies.

Alternating Currents	J. M. Gregory
Aspects of Isomerism	Peter Uzzell
Atomic and Molecular Weight Determination	R. B. Moyes
Catalysis in Chemistry	A. J. B. Robertson
Chemical Equilibrium	J. S. Coe
Collisions, Coalescence and Crystals	A. Ralph Morgan
Crystals and their Structures	I. F. Roberts
Energy in Chemistry	B. E. Dawson
Enzymes	Alan D. B. Malcolm
Fundamentals of Electrostatics	S. W. Hockey
The Inorganic Chemistry of the Non-Metals	John Emsley
Kinetics and Mechanisms of Reactions	B. E. Dawson
Logical Control Systems	Keith Morphew
The Mechanical Properties of Materials	R. A. Farrar
Nature Conservation	W. M. M. Baron
Oscillations	I. B. Hopley
The Polymorphism of Elements and Compounds	E. W. Jenkins
Radioactivity and the Life Sciences	D. J. Hornsey

Forthcoming

Biological Oceanography	M. V. Angel
Cell Biology	J. Chayen
An Introduction to Genetics	Michael A. Carter
Transition Metal Chemistry	J. J. Thompson
Waves	I. B. Hopley

532.5.1

24932

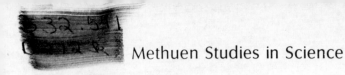

Methuen Studies in Science

Kinetics and Mechanisms of Reactions

B. E. DAWSON

Lecturer in Education
King's College, London

Methuen Educational Ltd

LONDON · TORONTO · SYDNEY · WELLINGTON

First published 1973
by Methuen Educational Ltd
11 New Fetter Lane, London EC4
© 1973 by B. E. Dawson
Printed in Great Britain by
William Clowes & Sons Ltd
London, Colchester and Beccles

ISBN 0 423 87510 8 non net
0 423 87520 5 net

All rights reserved.
No part of this publication
may be reproduced, stored in
a retrieval system, or transmitted
in any form or by any means,
electronic, mechanical, photocopying,
recording or otherwise, without the
prior permission of the publisher.

Distributed in the U.S.A. by
Harper & Row Publishers, Inc.
Barnes & Noble Import Division

Contents

249038 Oct 1975 B+T $3.50 PSC

Preface

An appreciation of chemical kinetics and its application to the study of reaction mechanism is of importance to all who wish to understand something of the 'dynamics' of Chemistry. At an introductory level, it is also appropriate to associate such studies with an appreciation of the nature of scientific inquiry and of the role of models in such inquiries. In addition, it is hoped that the approach used in this book places sufficient stress on the importance of relating theoretical studies with laboratory investigations.

This account of kinetics and mechanisms opens with an introductory survey of chemical processes and leads to an empirical treatment of reaction rates. From simple beginnings, a systematic approach to experimental measurements is introduced and then explored. The results of kinetic studies on specific chemical systems are then reviewed alongside those obtained by other means, thereby providing some justification for a use of specific reaction mechanisms. Thus far, the treatment provided requires a blending of theoretical and experimental aspects of the subject. Matters of a more abstract kind are discussed in the final chapter and include simplified treatments of both collision and transition state theories of reaction rates. No attempt is made to survey all aspects of kinetics and neither is it intended to provide a complete coverage of all aspects of the second theme for this book, reaction mechanism. It is hoped, however, that the selection provided will be appropriate for study at an introductory level and will serve as a stimulus for the student.

A general introduction of the kind provided by this book is highly dependent upon the published work of other authors. Indeed, my debt to authors of standard works, review articles and research papers is considerable and specific contributions are formally acknowledged in the text. As a teacher, it is also a pleasure to acknowledge the stimulation provided over the years by my friends and colleagues. In particular, I wish to thank Professor A. J. B. Robertson for commenting on the draft manuscript as a whole and Dr M. J. Perkins for helpful comments on the draft for chapter 3.

King's College, London
February, 1972   B.E.D.

1 Introduction

The study of chemical kinetics is concerned with the rates of chemical reactions, with all of the factors which influence the rates of specific reactions, and with the interpretation of reaction rates in terms of the detailed ways in which atoms, molecules or other chemical species behave during a reaction. The significance of the subject can be appreciated by referring to one of its more obvious applications. Chemists need to be able to establish the time required for carrying a reaction to completion and such information finds application in preparative work and in the design of industrial processes. However, the study of chemical kinetics is concerned with more than merely extending a list of facts about different chemical reactions. It is also concerned with the identification and subsequent use of fundamental principles which govern all chemical changes in our material world. On first acquaintance, there would appear to be very little to relate diverse processes such as the boiling of an egg, the fading of coloured materials used to furnish a home, the corrosion of metals, the combustion of petrol in the engine of a car, or the spectacular display of coloured light (and noise) which results from a use of fireworks. These changes and many other 'everyday' reactions seem somewhat remote from the chemical reactions employed in the production of metals from their ores, or in the manufacture of plastics and dyestuffs and all of the many other different types of material with which we are familiar. All of these reactions are so very different from each other and some are extremely difficult to describe in non-specialist language. Accordingly, if we wish to begin to understand something about the nature of chemical change and the associated rate processes, then we must attempt to visualise chemical processes in terms which are somewhat simpler to follow. We need to resort to the use of a 'model' situation in order to interpret a 'real' situation which may be quite different from our simple interpretation. Such a model will need modification to accommodate such knowledge and understanding as we have of the real situation. Indeed, the process of modification will be continuous until we reach the limit imposed by our thought processes or by the techniques employed in research work.

The use of models in science

A scientific model may be defined as a systematic analogy between a phenomenon whose properties and laws are well known and the event under systematic study. The function of such a model is to enable us to derive theoretical relationships which may then be tested on the real system. Quite often the results of experiments intended to test our ideas indicate that our interpretations can be but symbolic and

1

selective rather than specific. A model rarely provides a unique interpretation. In the case of the manner in which reactants are transformed into the products of reaction, it is often possible to devise several schemes to account for the more obvious facts about a given reaction. It follows that additional and more specific tests are then necessary to permit an investigator to distinguish between possible reaction pathways.

The course of a chemical reaction may be pictured in quite simple terms. First, the various reacting species may be visualised as coming together in such a manner that atoms from each of the species present can be re-grouped. This process must result in the breaking of some chemical bonds and the making of others. Secondly, once the products of reaction have been formed, they may well move away from the site of reaction. In the case of reaction between two simple molecular species, the process can be expressed in such terms visually. One can imagine the reaction taking place as a direct result of a 'favourable' collision by molecules. For example, the reaction between carbon monoxide and nitrogen dioxide under appropriate conditions can be expressed in such terms.

$$CO + NO_2 = CO_2 + NO$$

Similarly, the formation of iodoethane from hydrogen iodide and ethene (ethylene) in the gas phase can be considered in the same way.

$$HI + C_2H_4 = C_2H_5I$$

Indeed, these two reactions may be taken as representative systems which can be interpreted through a collision mechanism such as those shown in Fig. 1. This qualitative view appears to lend itself to development towards a better explanation of chemical reaction through a use of kinetic theory and experimental data.

Even so, this model may not easily be adapted to meet the requirements of every type of molecular behaviour. For example, the thermal decomposition of bromoethane can be represented by the equation

$$CH_3 \cdot CH_2 \cdot Br = CH_2 : CH_2 + HBr$$

and, for the present, it is somewhat more difficult to see just how one might adapt the systems shown in Fig. 1 to accommodate such behaviour. Clearly, some process is necessary to lead to a redistribution of energy within the molecular species bromoethane so that it will form new species, ethene (ethylene) and hydrogen bromide. The reaction is not and cannot be instantaneous: it occurs at a measurable rate and thereby provides some evidence on which to base a modification to our model system.

Yet other types of chemical change are possible. For example, electron transfer processes are legion. The reaction between iron(III) chloride and potassium iodide in solution provides a simple example of this type of reaction

$$2I^-(aq) + 2Fe^{3+}(aq) = I_2(aq) + 2Fe^{2+}(aq)$$

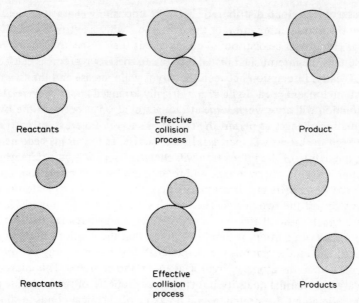

Fig. 1. Chemical reaction considered as a collision process: two possible 'model' processes

Now although this reaction is relatively simple to express in terms of a chemical equation, it is somewhat more difficult to interpret in terms of the simple collision model which we have used for other systems.

Any description of a chemical change requires a knowledge of the conditions which affect the rate of reaction, an appreciation of the nature of chemical bonds, and an appropriate consideration of the energetic and steric factors necessary to permit the reaction to occur. This suggests that a chemical reaction could well be made up of a number of separate stages or steps and this being so it will be necessary for us to acquire some knowledge of the rates at which each of these different steps in the overall process take place. This detailed knowledge will be necessary if one wishes to understand the factors controlling the rate of a reaction. In the most general terms, this type of description of a chemical reaction must apply to most, if not all, types of reaction including the simplest. By using a simple system as a model system, more complex reactions might then be considered by using the principles derived from the model. A convenient model system for us to consider is the formation and decomposition of a hydrogen molecule.

The hydrogen molecule

Imagine two hydrogen atoms set at some distance apart from one another and each consisting of one positively charged proton as nucleus and one negatively charged electron as a charge cloud distributed about the oppositely charged nucleus. In this condition, the mutual interaction of the two atoms will be extremely small and the two charge clouds will be distributed evenly about their respective nuclei. Now let us imagine that the internuclear distance between these two atoms is caused to decrease. Clearly, interaction between the atoms will increase and the two charge clouds will no longer appear to be symmetrically arranged about their related nuclei. Such a situation will arise when some attractive force exists between the two atoms. As the situation changes to permit the two atoms to get closer, so this attractive force between the atoms will increase. In due course, as the atoms become closer together, the nature of the interaction will change. There will be a time when quite a large repulsive force will overcome and replace the force of attraction. This situation can arise when there is a direct interaction of the two electron clouds and, at an even later stage, of the two nuclei. The overall situation may be represented graphically as in Fig. 2 which shows the variation in potential energy between the two atoms with internuclear separation. It should be noted that the depth of the energy minimum shown on the graph is the characteristic change in energy per molecule which occurs when the atoms of hydrogen are said to combine. This energy minimum is *not* an average thermal quantity but is a direct property of the molecule concerned.

Information about the stretching and breaking of chemical bonds in simple diatomic molecules can be obtained from spectroscopic studies. In the case of hydrogen, it can be shown that an energy of about 435 kJ is required to break all of the bonds present in one mole of hydrogen gas.

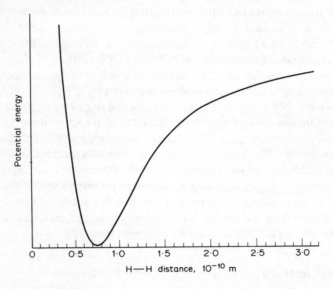

Fig. 2. The variation of potential energy with distance between the atoms in the diatomic molecule, H_2

It should be noted that Fig. 2 also relates to the formation of a hydrogen molecule. The minimum in the potential energy curve occurs when the internuclear separation of the atoms is 0.74×10^{-10} m (0.074 nm). In addition, if the atoms in a molecule of hydrogen were held at a separation of, say 1×10^{-10} m and then released, the molecule would not break but begin to vibrate. The internuclear distance would be seen to decrease at first until it was a little less than 0.74×10^{-10} m and then increase again to 1×10^{-10} m. This type of vibration would continue until such time as some other disturbance might affect the molecule in question. The frequency of these vibrations is about 10^{13} Hz, the precise value being dependent upon the exact shape of the potential energy curve. It is through a careful study of vibrational frequencies exhibited by molecules (acquired from spectroscopic data) that we obtain details of the exact shapes of potential energy curves.

The shape of a potential energy curve also provides information leading to the determination of the amount of energy necessary to break a particular bond (the bond dissociation energy). It may be recalled that this quantity of energy depends not only on the nature of the bond but also on its structural environment. For example, three atoms constituting a molecule of water may be considered to be held together by two OH-bonds. It is noteworthy, however, that the dissociation of a water molecule into its constituent atoms requires two different and separate energetic processes

$$H_2O \rightarrow H + OH$$

$$OH \rightarrow O + H$$

The energy required to bring about the first of these two dissociations is about 494 kJ mol^{-1} whereas that needed for the second is only about 418 kJ mol^{-1}. In both instances, the quantity of energy required relates to the use of one mole of starting material: that is of water in the case of the first of these two reactions and of hydroxyl radicals in the second. Thus, these two OH-bonds possess different dissociation energies because of structural environmental differences between the species concerned, a water molecule and a hydroxyl radical. A similar situation will obtain for the bond dissociation energies of the OH-bonds in methanol and phenol.*

A simple exchange reaction

A simple exchange reaction of the type

$$X + YZ \rightarrow XY + Z$$

provides a natural sequel to the foregoing discussion about the hydrogen molecule if the symbols X, Y and Z relate to separate species. In the simplest case, where the

* Note: a discussion making the distinction between 'bond energies' and 'bond dissociation energies' appears in Dawson, B. E. *Energy in chemistry: an approach to thermodynamics.* Methuen, 1971 (Chapter 2).

symbols actually refer to separate atoms of hydrogen, the reaction will allow a formal study to be made of the process of making one covalent bond and of breaking another. In effect, such a system would be the simplest possible extension in mechanistic terms of the process already reviewed. The reaction process might be considered to begin when the atom X moves towards the molecule YZ such that little or no interaction between the species occurs. This stage of the reaction could be represented by

$$X \quad Y-Z$$

As the gap between the two species is narrowed, interaction will gradually increase in much the same manner as when two atoms of hydrogen were allowed to 'react'. At some stage in this process, the atom Y will appear to be shared by both atom X and atom Z and the situation could be represented by

$$X..Y..Z$$

before the atom Z moves away from Y thereby permitting the formation of the molecule XY, as indicated by

$$X-Y \quad Z$$

To provide a more exact deduction, we need to consider a definite geometrical arrangement of atoms. Thus, if the three atoms are arranged in a straight line, as shown above, we can represent the potential energy changes during such a reaction in a manner not unlike that normally adopted to display contours on a map. Thus, the distance between the atom X and the atom Y during the entire process could be plotted along the Y-axis of a graph (as R_{xy}) and the distance between the atom Y and the atom Z along the X-axis (as R_{yz}). All points of equal potential energy might then be plotted and joined up in the same manner as that employed to display contours. Thus, where the lines of equal potential energy are shown to be drawn close together, a reacting species (such as the atoms X, Y or Z) would experience rapid changes in potential energy when caused to move across such contours. (This is feasible since there is an energy transformation from kinetic to potential energy.) Similarly, when contours of equal potential energy are spaced relatively far apart, a less steep energy surface may be considered to exist.

Fig. 3 provides such representations of a reaction: a three-dimensional 'model' and a simpler two-dimensional representation. In both, the point a, for example, represents a configuration where the atom X is far removed from the molecule YZ; b appears on an energy plateau such that the three atoms X, Y and Z are well separated; c corresponds to the situation where atom Z is far removed from molecule XY; and d relates to a zone on the contour map where all three atoms are extremely close together as formulated by the representation $X..Y..Z$. This special zone d might relate equally well to the effective collision of species X with species YZ or to Z with XY.

6

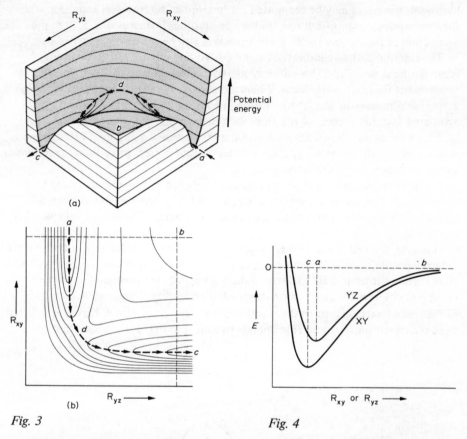

Fig. 3

Fig. 4

Fig. 4 shows two cross-sections of the energy contours indicated on the two-dimensional 'map' in Fig. 3. The curve YZ is a cross-section of the potential energy surface drawn about the line *ab* shown in Fig. 3. The curve marked XY relates to the surface drawn through *bc*. It should be noted that in the interest of clarity, it has been assumed that the depth of the energy valley about *c* is somewhat greater than that about *a*. Using this interpretation, it requires but little imagination to appreciate that these two energy valleys meet at the pass shown in Fig. 3 as *d*.

It follows from this interpretation that the reaction represented by

$$X + YZ \rightarrow XY + Z$$

corresponds to the path leading from valley *a* through the zone *d* into the valley *c*. Perhaps the most significant point to note is that *d* relates to an increase in potential energy, a matter which can be accounted for in terms of the repulsive force exerted when atom X approaches molecule YZ (or when atom Z approaches the molecule XY). In the case of the exchange reaction between an atom and a molecule of

7

hydrogen, the effect may be accounted for in terms of internuclear repulsion when the reacting species approach one another. In all other instances of reaction, the interaction of filled inner shells of electrons will also be relevant.

The reaction pathway indicated by the dotted line joining a, d and c in Fig. 3 is not a unique situation. This pathway merely corresponds to the *least* energy requirement for reaction to occur. Where a diatomic molecule taking part in such a reaction processes an oscillatory motion in addition to the translational motion associated with the process of reaction, then the reaction pathway might take the form of a zigzag line drawn across, for example, the valley a and then through the zone d and on into the valley c. Such a system would possess more than the minimum energy necessary for reaction to occur.

It will be appreciated that not all collisions between reacting species need lead to 'reaction'. This behaviour might be accounted for by representing the reaction pathway as a line drawn about the upper portion of one of the energy valleys, such as a in Fig. 3.

However, the real virtue of this simple representation of a chemical reaction rests with the possibility of obtaining a cross-section of an actual reaction pathway. The cross-section shown in Fig. 5 relates to the least energy requirement for reaction to occur and observes the same conventions adopted for Fig. 3 and Fig. 4. The state of the species participating in the reaction at the top of the energy barrier is referred to as the *transition state* and corresponds to zone d in Fig. 3. It follows that the

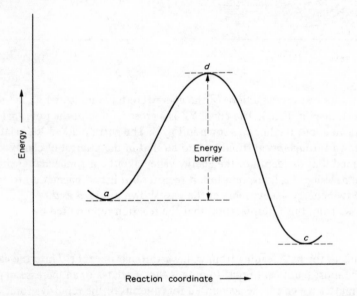

Fig. 5. A cross-section of the reaction pathway a . . . d . . . c shown on Fig. 3 where a corresponds to reactants, d the transition state and c the products of reaction

transition state can exist only as a direct result of an effective collision between reacting species and it will decompose as soon as vibrations occur in a definite mode in one or more of the bonds present in this complex. In the case of the reaction between a hydrogen molecule and a hydrogen atom, a recent calculation (1968) indicates that the height of the barrier relative to the energies of the reactants is about 47.5 kJ mol^{-1}.

Other reactions

The reaction systems outlined so far have been extremely simple and only required a knowledge of the covalent bond. Other systems may employ more than three atoms and may also require an appreciation of other types of chemical bond. To understand such systems will require more knowledge than has been assumed so far. However, at this stage, it is unnecessary to test and modify our model of a chemical system against every variation if we are content merely to apply the ideas to other systems without seeking a formal proof in each instance. (Thus, if one observed a reaction system of more than three atoms, the number of co-ordinates required to fix their relative configurations becomes too large to visualise an appropriate potential energy surface.)

The discussion of a simple exchange reaction in the previous section has also implied an additional simplification, namely the existence of a unique potential energy surface for that reaction. In practice, each reaction system will possess a large number of such surfaces corresponding to different electronic states of the species taking part in the process. Our considerations relate to the lowest state only and in many instances this is the most important. A more detailed exposition dealing with 'excited states' would be inappropriate at this stage.

Fig. 6 shows two general situations which will be familiar to all who have worked in an elementary chemistry laboratory. In both instances, the activation energy

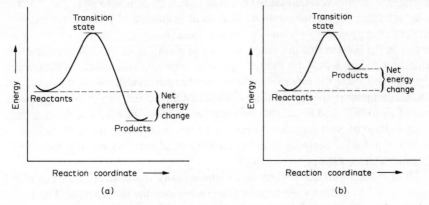

Fig. 6. Activation Energy as an energy barrier between reactants and products: (a) exothermic reaction; (b) endothermic reaction

9

necessary for reaction to occur is shown as an energy barrier between reactants and products. Where the products of reaction are energetically more stable than the reactants, the situation relates to that obtaining for an *exothermic reaction.* The reverse situation is also shown and refers to the net energy change for an *endothermic reaction.* These reaction pathway profiles are of the simplest possible kind. If one had more complex profiles with several maxima and minima, then the same basic ideas would still apply to each portion of the curve. Thus, an energy barrier would still separate the reactants from the products for each step and the highest point on each portion of the energy profile would still correspond to the transition state of the step portrayed. The value of using this type of profile is deferred until Chapter 3 where basic principles are reviewed and applied to the study of specific reactions.

Reaction mechanism from a study of reaction kinetics

The study of a particular reaction using the procedures of chemical kinetics can supply experimental evidence which may then be interpreted in terms of the detailed ways in which atoms, molecules or other chemical species behave during a reaction. This behaviour is often referred to as a reaction mechanism. However, it is as well to remember that such mechanisms are merely interpretations or hypotheses which need to be tested by further investigation. Indeed as new facts are discovered about a reaction, an accepted mechanism needs to be revised. It is equally important to remember that a given set of empirical facts can be interpreted on occasion by more than one mechanism!

Kinetic studies are often preceded by preliminary investigations for these can supply the means whereby reactants and products can be identified and distinguished from one another. The stoichiometry of the reaction can also follow from such studies and enables a chemist to account for the use of the reactants in producing the products of reaction. It can also lead to some understanding of the conditions necessary to ensure reaction taking place or to test its completeness.

By selecting the various methods of analysis employed to identify various reactants and products, it is usually feasible to estimate the quantity of one or more species in the presence of the others with some good degree of accuracy. The experimental procedure selected is then tested on various synthetic mixtures before being applied to estimate the rate of reaction under known conditions. In this way, it is possible to measure the rate of disappearance of a reactant, or the rate of appearance of a product, and to use this information to assess the effect of varying the concentration of each reactant in turn on the reaction rate. Other tests include the use of potential catalysts to influence the rate of reaction and also the effect of temperature on the rate.

All of this information is then used systematically to derive an empirical rate equation and to deduce various potential mechanisms for the reaction. The analysis employed must account for all of the known facts about the reaction and predict effects to be assessed by separate experimental studies. The overall result can lead to a detailed knowledge of the reaction in terms of a reaction mechanism. On occasion, this result can be assessed by other methods, such as isotopic analysis, which are not dependent upon the procedures used in rate studies, (cf. Chapter 3). However, it is only when information about a given mechanism is supported by a variety of separate investigations that it can be used with confidence. This must be remembered whenever attempts are made to account for reactions. Fortunately, a great deal can be achieved by using simple investigatory studies of the type described in the next chapter.

2 Rate measurements

Most people might accept the fact that chemical reactions can occur under a variety of very different conditions and can proceed at very different rates. The rate of a given reaction must depend on the 'availability' of the reactants, on the presence or absence of catalysts, and on the temperature. It remains for us to identify the precise relationship between these variables and to relate our findings to molecular events so as to identify the steps of an actual mechanism. To assess the effect of any one factor on the rate of a reaction, it would seem reasonable to arrange for all other variables to be held constant. This may seem obvious but on occasion it can prove extremely difficult to achieve in practice. For example, methyl formate (methyl methanoate), a simple ester, can be hydrolysed under a variety of conditions to form formic acid (methanoic acid) and methanol. The overall reaction can be represented by the equation

$$H.CO_2.CH_3 + H_2O = H.CO_2.H + CH_3.OH$$

Now, if the reaction is conducted in an acidic solution at constant temperature such that the amount of acid present is in great excess, it is possible to measure the rate of disappearance of methyl formate in terms of the rate of formation of formic acid.

$$H.CO_2.CH_3 \xrightarrow[\substack{\text{excess} \\ \text{aqueous} \\ \text{acid}}]{} H.CO_2.H$$

The only variable involving the concentrations of reactants in such a system is dependent upon the initial concentration of the methyl formate used. So, if, for example, 10 cm^3 methyl formate had been added to 200 cm^3 of approximately 0·5M hydrochloric acid in a vessel held at constant temperature and 2 cm^3 aliquots (i.e. samples of the reaction mixture) removed and quenched at known times in 100 cm^3 of ice-cold water, then by titrating each diluted solution of the reaction mixture with 0·10M sodium hydroxide in the usual way some measure of the extent of reaction can be acquired. (It will be appreciated that the purpose of transferring an aliquot of the reaction mixture to a large volume of ice-cold water is to effectively stop the reaction proceeding any further in that sample. Accordingly, the composition of the reaction mixture can be estimated at known intervals of time.) Some typical results from one such experiment are given in Table 1.

The effect of the concentration of hydrochloric acid on the rate of reaction at a given temperature can be determined by a series of separate experiments (how?).

The stoichiometric equation for the hydrolysis shows that one molecule of formic acid is formed from one molecule of methyl formate. Accordingly, the concentrations of both methyl formate and formic acid can be calculated at specific instances from the titres shown in column 2 of Table 1, after making due allowance for the hydrochloric acid present in the reaction mixture. The formation of formic

Table 1 *Hydrolysis of methyl formate in acid solution at 18°C*
(after R. A. Bowman, 1945)

time min	0·10M NaOH cm^3	[formic acid] expressed as cm^3 0·10M NaOH	[methyl formate] expressed as cm^3 0·10M NaOH
0	11·3	0	12·0
10	15·5	4·2	7·8
20	17·5	6·2	5·8
30	19·6	8·3	3·7
40	20·8	9·5	2·5
50	21·5	10·2	1·8
60	22·1	10·8	1·2
70	22·6	11·3	0·7
100	23·0	11·7	0·3
180	23·3	12·0	0
240	23·3	12·0	0
∞	23·3	12·0	0

acid during the hydrolysis can be displayed in the manner shown in Fig. 7. The disappearance of methyl formate during the reaction can be displayed in a similar manner (Fig. 8). Both of these representations demonstrate that the rate of reaction is dependent upon the concentration of methyl formate at a given instant. Thus, the slope of the curve shown in Fig. 8 can be estimated by using a simple plane mirror to draw tangents at specific points on that curve. Fig. 9 uses such results and shows precisely the dependence of the reaction rate at any instant on the concentration of methyl formate at that time.

Fig. 10 shows three separate sets of results for experiments performed under identical conditions, save for the initial concentration of methyl formate. It is instructive to compare these decay curves and thereby determine their characteristics. The decay curve for quite a different system is displayed in Fig. 11 and it can be seen that this curve possesses quite different characteristics. The curve displayed in Fig. 11 clearly depends in some way on the initial concentration of the reactant whereas such a dependence is not shown by the curves in Fig. 10.

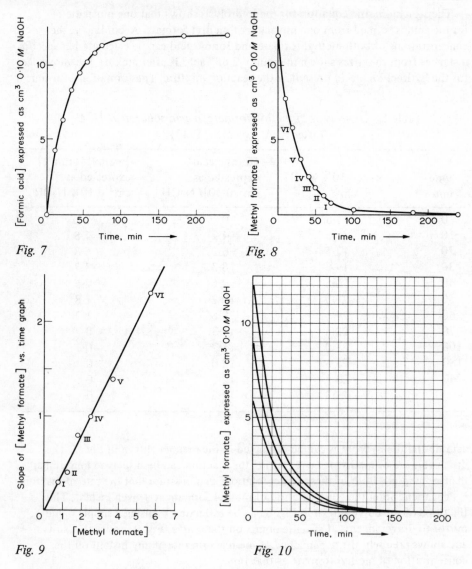

Fig. 7

Fig. 8

Fig. 9

Fig. 10

EXPERIMENT (1) Prepare a solution of potassium iodide in water and add to it a dilute solution of hydrogen peroxide. Investigate the effect of adding dilute sulphuric acid to this mixture.

(2) Mix together a small volume of acetone and a few drops of iodine solution. Investigate the effect of adding dilute hydrochloric acid to this mixture.

(3) Using small quantities of reagents, compare the reactions of silver nitrate in aqueous-ethanol with various haloalkanes.

14

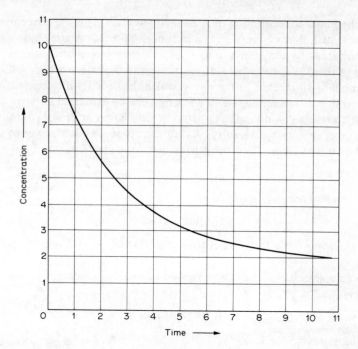

Fig. 11

(4) Prepare a dilute solution of sodium dichromate in 4*M* sulphuric acid and use it to investigate the chromic acid oxidation of substituted ethanols.

Order of Reaction

The results from the four qualitative investigations contrast sharply with those displayed in Fig. 7, 8 and 9 for the hydrolysis of methyl formate in hydrochloric acid solution. Each of the systems studied yields distinctive results which do not appear to relate in any way with those from another system. It is apparent that we require some more formal method of presenting the results of investigation and to consider the possibility of presenting quantitative data graphically.

The hydrolysis of methyl formate results in the formation of formic acid and, in the most general terms, we could represent such a reaction by

$$A \longrightarrow B$$

(methyl formate)   (formic acid)

We could summarise the experience gained from the analysis of the results shown in Table 1 by

$$-d[A]/dt = d[B]/dt = dx/dt = k(a - x) \qquad (2.1)$$

15

where a is the initial concentration of A and the term $(a - x)$ is the concentration of A at time t, such that x is the decrease in concentration of A after time t. The constant term k can be considered to be a proportionality constant with the dimensions of time^{-1}. The symbol $-\mathrm{d}[A]/\mathrm{d}t$ refers to the rate at which A disappears and the symbol $\mathrm{d}[B]/\mathrm{d}t$ refers to the rate at which the product B is formed.

In other instances where the rate of a reaction can be expressed in terms of the rate at which a reactant A disappears, $-\mathrm{d}[A]/\mathrm{d}t$ may be proportional to the concentration of species A, $[A]$, or to $[A]^2$ or even to some other power of the concentration of species A. Thus, the general case for a reaction

$$A \rightarrow \text{products}$$

the rate of reaction may be expressed as

$$-\mathrm{d}[A]/\mathrm{d}t = k[A]^{\alpha}$$

where the exponent α is referred to as the order of reaction with respect to the reactant A.

A more complex change can be represented in a similar manner. For example, the reaction

$$A + B + C \rightarrow \text{products}$$

may proceed at a rate r, such that

$$r = k\,C_A{}^{\alpha}C_B{}^{\beta}C_C{}^{\gamma},$$

where k is a proportionality constant, C_A the concentration of species A, C_B the concentration of species B, C_C the concentration of species C, the exponent α the order of reaction with respect to species A, the exponent β the order of reaction with respect to species B, and the exponent γ the order of reaction with respect to species C. When the rate of reaction can be expressed in this form, it is worth noting that the exponents α, β and γ are either small integers or fractions.

A reaction could be even more complex, as with

$$pA + qB + rC \rightarrow \text{products}$$

and could be governed by a rate equation of the form

$$r = \mathrm{d}x/\mathrm{d}t = k(a - x)^{\alpha}(b - qx/p)^{\beta}(c - rx/p)^{\gamma} \qquad (2.2)$$

where a is the initial concentration of A, b the initial concentration of B, and c the initial concentration of C. The term x is the decrease in concentration of A after time t and so $\mathrm{d}x/\mathrm{d}t$ is the measure of the rate of reaction. As before, the exponents α, β and γ are small integers or fractions and are orders of reaction with respect to the reactants A, B and C, respectively. The sum of the exponents $(\alpha + \beta + \gamma)$ is referred to as the *overall* order of reaction. The coefficients q/p and r/p are determined by the stoichiometry of the reaction. Again the constant k is a proportionality constant known as the rate constant. The dimensions of this constant will clearly depend on the number of concentration terms in the rate equation.

General rate equation for first order reactions

The rate of a first order reaction is directly proportional to the concentration of one of the reacting species only. Thus for the reaction

$$A \rightarrow products$$

the rate equation is

$$dx/dt = k(a - x). \tag{2.3}$$

The constant k used in (2.3) is known as the first order rate constant. Rewriting (2.3) in the form

$$dx/(a - x) = k \, dt$$

and integrating, yields an equation from which a value for the first order rate constant can be obtained. Thus,

$$-\ln(a - x) = kt + c$$

where c is a constant. To find the value of c, we require related values of t and x. Thus when $t = 0$, x must also be zero and so c must have the value $-\ln a$. Hence, we can write

$$kt = \ln a - \ln(a - x). \tag{2.4}$$

On re-arranging, this equation yields k in terms of a, x and t:

$$k = \ln\{a/(a - x)\}/t$$

$$= 2 \cdot 303 \, \lg\{a/(a - x)\}/t. \tag{2.5}$$

This last equation may be written in the form of the general equation for a straight line, $y = mx + c$:

$$\lg(a - x) = -(k/2 \cdot 303)t + \lg a, \tag{2.6}$$

which allows some appreciation of graphical methods of data presentation for first order reactions, (cf. Fig. 12 and Fig. 13).

General rate equation for second order reactions

The rate of a second order reaction is proportional to the product of two concentration terms. For the general case of a reaction between two species A and B, such that

$$A + B \rightarrow products.$$

The rate of reaction will be given by

$$dx/dt = k(a - x)(b - x),$$ (2.7)

since the coefficients and exponents are equal to unity (cf. (2.2)). In equation (2.7), the rate constant k will be a second order rate constant. In certain cases, equation (2.7) can be simplified. Thus when $a = b$ or when the species A and B are identical, equation (2.7) becomes

$$dx/dt = k(a - x)^2,$$

i.e.

$$dx/(a - x)^2 = k \, dt.$$

On integrating, this becomes

$$1/(a - x) = kt + c$$

where c is a constant. The value of c can be found by using related values of t and x. Thus, when $t = 0$, $x = 0$ and so c must equal $1/a$. Hence, we can write

$$1/(a - x) = kt + 1/a$$ (2.8)

from which it can be seen that the second order rate constant, k, will possess units of reciprocal time and reciprocal concentration.

These derivations suggest that equation (2.2) can be used to obtain other general rate equations. However, as will become apparent, in practice it is found to be more useful to consider the rate equation for a system under investigation rather than confine discussion to general derivations and considerations.

Some second order reactions

In this section, the reader is invited to consider a small number of relatively simple reactions in appropriate detail and to derive a rate equation for each of these systems using the same general principles.

The hydrolysis of ethyl acetate (ethyl ethanoate) under alkaline conditions   This reaction can be represented by the following equation

$$CH_3 . CO_2 . C_2H_5 + OH^- = CH_3 . CO_2^- + C_2H_5 . OH$$

It can be shown by experiment that this reaction is first order with respect to ethyl acetate and first order with respect to hydroxide ions. The overall order of reaction is two. If we represent the initial concentration of ethyl acetate by a and the initial concentration of hydroxide ions by b, and if the decrease in concentration of the reactants after time t is x, then the concentrations of the reactants at time t will be $(a - x)$ in the case of ethyl acetate and $(b - x)$ in the case of hydroxide ions. It follows that the rate equation for this reaction will be given by

$$dx/dt = k(a - x)(b - x) \qquad (2.9)$$

from which it follows that the rate constant for this reaction will be given by

$$k = \frac{2 \cdot 303}{(b - a)t} \lg \frac{a(b - x)}{b(a - x)}. \qquad (2.10)$$

The hydrolysis of ethyl acetate (ethyl ethanoate) under acid conditions   The equation for this reaction can be written in the form

$$CH_3 . CO_2 . C_2H_5 + H^+(aq) + H_2O = CH_3 . CO_2H + C_2H_5 . OH + H^+(aq)$$

The reaction is catalysed by hydrogen ions. It can be shown that the reaction is first order with respect to ethyl acetate and first order with respect to hydrogen ions. It has an overall order of reaction of two. If we represent the initial concentration of ethyl acetate by a and the initial concentration of hydrogen ions by b, then after time t the concentration of ethyl acetate will be $(a - x)$, where x is the decrease in concentration. It follows that the equation for the reaction is

$$dx/dt = k(a - x)b \qquad (2.11)$$

where k is a second order rate constant. It can be shown that the integrated form of this equation is

$$k = \frac{2 \cdot 303}{tb} \lg \frac{a}{(a - x)}, \qquad (2.12)$$

which is very similar in form to equation (2.5). However, it should be noted that the rate constants in these two equations have different dimensions. (Why?) It should also be noted that the concentration of the hydrogen ions remains constant throughout this hydrolysis. Accordingly, the system will show first order kinetics for a given concentration of hydrogen ions.

Experiment   How could one show that the concentration of hydrogen ions affects the rate of reaction during the acid hydrolysis of a simple ester? Devise a set of quantitative experiments to determine the second order rate constant for the hydrolysis of methyl formate under acid conditions at a constant temperature.

The iodination of acetone (propan-2-one) under acid conditions   The overall reaction for this process can be written in the form

$$CH_3 . CO . CH_3 + H^+ + I_2 = CH_2I . CO . CH_3 + 2H^+ + I^-$$

The reaction can be shown to be first order with respect to acetone, first order with respect to hydrogen ions, and zero order with respect to iodine. This means that the concentration of iodine used does not influence the rate of this reaction

19

in any way. If we represent the initial concentration of acetone by a and the initial concentration of hydrogen ions by b, then at time t the concentration of acetone will be $(a - x)$ and the concentration of hydrogen ions will be $(b + x)$, because of the stoichiometry of the reaction. Accordingly, the rate equation for this reaction will be given by

$$\mathrm{d}x/\mathrm{d}t = k(a - x)(b + x). \tag{2.13}$$

The integrated form of this equation gives the rate constant for this reaction

$$k = \frac{2 \cdot 303}{(a + b)t} \lg \frac{a(b + x)}{b(a - x)}. \tag{2.14}$$

The oxidation of iodide ions by persulphate ions (peroxodisulphate ions)   The stoichiometric equation for this reaction is

$$S_2O_8{}^{2-}(aq) + 2I^-(aq) = 2SO_4{}^{2-}(aq) + I_2(aq)$$

It can be shown that the reaction is first order with respect to persulphate ions and first order with respect to iodide ions. If the initial concentration of persulphate ions were a and the initial concentration of iodide ions were b, then after time t the respective concentrations of the reactants will be $(a - x)$ and $(b - 2x)$ due to the stoichiometry of the reaction. It follows that the rate equation for this system will be

$$\mathrm{d}x/\mathrm{d}t = k(a - x)(b - 2x) \tag{2.15}$$

where x and k possess the usual meanings. The rate constant k is obtained from the integrated form of (2.15) and has the value

$$k = \frac{2 \cdot 303}{(b - 2a)t} \lg \frac{a(b - 2x)}{b(a - x)}. \tag{2.16}$$

Not all second order reactions exhibit unique rate equations. Indeed many simple systems obey the simplest form of second order rate equation (i.e. equation (2.8)). These include gas phase decomposition reactions like

$$2HI \rightarrow H_2 + I_2$$

and gas phase dimerizations like

$$2CH_2{:}CH . CH{:}CH_2 \rightarrow \begin{array}{c} CH_2 \\ HC \diagup \quad \diagdown CH . CH{:}CH_2 \\ \| \qquad \quad | \\ HC \diagdown \quad \diagup CH_2 \\ CH_2 \end{array}$$

The purpose of the four special examples considered in this section is to stress the importance of making use of actual experimental data rather than merely applying general rules to a chemical system.

20

Some first order reactions

Illustrative examples of systems which obey first order kinetics include decomposition and isomerisation reactions in the gas phase, and a variety of reactions in solution. For example, the decomposition of di-t-butyl peroxide into acetone and ethane obeys first order kinetics:

$$(CH_3)_3C.OO.C(CH_3)_3 \rightarrow 2CH_3.CO.CH_3 + CH_3.CH_3$$

The decomposition of cyclobutane and the isomerization of cyclopropane in the gas phase follow the same rate law at pressures above one atmosphere

$$\begin{array}{c} H_2C-CH_2 \\ | \quad\ | \\ H_2C-CH_2 \end{array} \rightarrow \quad 2CH_2{:}CH_2$$

$$\begin{array}{c} H_2C-CH_2 \\ \diagdown\ \diagup \\ CH_2 \end{array} \rightarrow CH_3.CH{:}CH_2$$

The hydrolysis of 2-iodo-2-methylpropane in aqueous alcohol also observes first order kinetics.

$$(CH_3)_3C.I + HOH \rightarrow (CH_3)_3C.OH + HI$$

The foregoing discussions emphasise the empirical nature of orders of reaction. The determination of orders of reaction forms an important part of any investigation of a reaction mechanism using the techniques of reaction kinetics. In many instances, a basis for the interpretation of experimental results expressed in terms of mechanism can begin from a simple comparison of the empirical rate equation with the stoichiometric equation for the reaction. In instances where reactions do not possess a simple integral order of reaction, we may conclude that the mechanism of the reaction is complex. On occasion, experimental results can be deceptively simple and misleading. Indeed, all interpretations require careful assessment. Any determination of the number of specific molecules or other particles which meet in reactive collision processes depends on an interpretation of the available evidence. We must regard all such elementary processes which make up a mechanism as tentative suggestions. Experience shows that an elementary process involving two chemical species (known as a bimolecular process) is usually second order. Similarly, a process involving three chemical species (a termolecular process) is usually third order. However, it is worth recording that the reverse of these two statements is less likely to be true, (cf. Chapter 3).

Half-life of reactions

It is convenient to return to the discussion based on Fig. 10 and Fig. 11 and to make use of the mathematical analysis on pages 15 to 20. It is not difficult to show that the time required to complete a definite fraction of the reaction (e.g. one-half, one-tenth, etc.,) is independent of the initial concentration of the reactant for a process which follows first order kinetics whereas for a second order reaction with 1:1 stoichiometry (i.e. the simplest case) it is inversely proportional to the initial molar concentration of the reactants if both are the same.

PROBLEM   Test the above suggestion on equations (2.5) and (2.8).

In general terms, this result may be expressed in the form

$$\tau \propto 1/a^{n-1} \tag{2.17}$$

where τ is the time required to complete a certain fraction of the reaction, a the initial concentration of reactants and n the order of reaction.

PROBLEM   Use (2.17) to determine the order of reaction for each system shown in Fig. 10 and Fig. 11.

If in two different experiments the initial concentrations of the reactants are a_1 and a_2 and the corresponding times of reaction are τ_1 and τ_2 respectively, it follows from (2.17) that

$$\frac{\tau_1}{\tau_2} = \left(\frac{a_2}{a_1}\right)^{n-1}, \tag{2.18}$$

and so

$$n = 1 + \frac{\lg(\tau_1/\tau_2)}{\lg(a_2/a_1)}. \tag{2.19}$$

It is often convenient to adopt the half-life of a reaction as a unit of time when comparing data—such as shown in Fig. 10 and Fig. 11. The determination of the half-life of a reaction is fairly simple and can be used to obtain the rate constant. Thus, for a first order reaction, one can use (2.5) to show

$$k = (2 \cdot 303 \lg 2)/t_{\frac{1}{2}} \tag{2.20}$$

and for a second order reaction

$$k = 1/(a t_{\frac{1}{2}}) \tag{2.21}$$

in the simplest case (cf. equation (2.8).

EXPERIMENT – DISCUSSION   The stoichiometric equation for the reaction between iodide ions and hydrogen peroxide under acid conditions is

$$H_2O_2 + 2I^- + 2H^+ = I_2 + 2H_2O$$

It is suggested that the reader investigates the kinetics of this system and determines the rate constant at one temperature.

22

Preliminary studies on this system (cf. p. 14) will reveal that iodide ions and hydrogen peroxide react somewhat slowly in the absence of hydrogen ions; that the addition of acid increases the rate of reaction; and that starch can be used to detect the formation of iodine. Clearly, it will be essential to make a series of experiments on this system if a full investigation is to follow. By using reaction mixtures where the concentrations of hydrogen ions and iodide ions are held constant, the rate of disappearance of hydrogen peroxide could be assessed from the rate of appearance of an equivalent quantity of iodine at a given temperature. By including some starch in the iodide solution, by using an excess of iodide and of acid, these conditions could be met. In addition, it will be found convenient to add to the original reaction mixture a small but known quantity of sodium thiosulphate solution (e.g. 2 cm^3 0·10M). Under these conditions, the time taken to change the colour of the reaction mixture must be proportional to the amount of thiosulphate used (if one assumes, quite correctly, that the reaction between iodine and thiosulphate ions is extremely fast when compared with the system under investigation). So, by adding equal volumes of sodium thiosulphate solution to the reaction mixture at intervals when the starch indicator changes colour and noting the times of addition, the concentration of iodide ions can be held constant and the concentration of hydrogen peroxide can be expressed in terms of the volume of sodium thiosulphate solution used.

Procedure: Prepare the following solutions—

A Dilute a mixture of 50 cm^3 0·10M potassium iodide and 10 cm^3 1% starch solution to 250 cm^3. Transfer this solution to a large flask and allow it to equilibrate in a thermostat.

B Dilute a mixture of 100 cm^3 3·5M sulphuric acid and 25 cm^3 0·05M hydrogen peroxide to 250 cm^3. Allow the solution to equilibrate in a thermostat.

C Prepare a 0·10M solution of sodium thiosulphate.

Experiment 1: Add 2 cm^3 of C to A. Add B to A and note the time of half addition. Add further 2 cm^3 of C to the reaction mixture when the mixture changes colour and record the times of these additions. From the data obtained, estimate the half-life of the reaction.

Experiment 2: Repeat the first experiment but use 50 cm^3 0·05M hydrogen peroxide in place of 25 cm^3 when making up solution A. From the data obtained, estimate the half-life of the reaction.

Experiment 3: Repeat the first experiment using 25 cm^3 0·10M potassium iodide in place of 50 cm^3 when making up solution A. Use your results to determine the order of reaction with respect to iodide ions.

Finally, use the results of these three experiments and the preliminary investigations (cf. p. 14) to write the rate equation for this reaction. Can you use this result to deduce a mechanism for the reaction?

Presentation of experimental results

The human eye is surprisingly good at judging the optimum straight line through a set of points provided all of the points carry equal weight. The analysis given on page 15 et seq. provides the means of obtaining linear functions for the presentation of kinetic data. As will become apparent, the use of both first and second order plots for the same data stresses the importance of following a reaction until it is almost complete. Equations (2.6) and (2.8) provide the basis for most linear presentations of rate data. Fig. 12 and Fig. 13 depend on equation (2.6) and show the data given in Table 1 for the hydrolysis of methyl formate under acid conditions at 18°C. Fig. 14 depends on equation (2.8) and employs the same basic data.

Both forms of first order plot find use in the presentation of experimental results. That used in Fig. 14 provides, in this instance, a useful confirmation of the result shown in Fig. 13. (Why?). To make use of equation (2.8) and the form of presentation shown in Fig. 14 for second order reactions, the initial concentrations of both

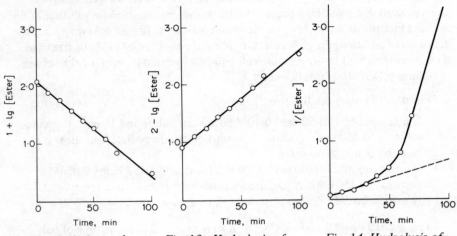

Fig. 12. Hydrolysis of methyl formate: first order plot

Fig. 13. Hydrolysis of methyl formate: first order plot

Fig. 14. Hydrolysis of methyl formate: second order plot

reactants must be equal. If the second reactant has a concentration higher than the first, then the graph—when based on the relationship shown by Fig. 14—will show upward curvature. Similarly, if the concentration of the second reactant is less than the first, a downward curvature will be exhibited. A number of other factors can lead to 'curvature' effects of this type and include changes in temperature in the reaction mixture and changes in the reaction medium in those cases where reactions are followed in relatively concentrated solutions. Other effects can be traced to the presence of impurities and to side reactions. It follows that it is often wise to repeat a given investigation under set conditions before attempting to interpret

24

results. Indeed, all such work must allow for the two chief causes of unreproducibility: human error and unsuspected variation in the starting materials used in experiments. It is equally important not to restrict the interpretation of experimental results to a single method, especially when determining the order of reaction with respect to a reactant.

Problem Use the information shown in Fig. 11 to obtain a second order linear plot. What happens if the data is used on a first order plot?

The influence of temperature on the rate of a reaction

A great deal of elementary experience of chemical reactions supports the view that for most chemical systems an increase in temperature results in an increase in the rate of reaction. Indeed, if a given reaction is carried out at a series of temperatures, it will be found that in most instances the reaction rate will increase in a regular manner. Fig 15(a) illustrates this relationship by showing the effect of temperature on the second order rate constant for the thermal decomposition of hydrogen iodide. In other instances, such as when a change in reaction mechanism occurs or when some other effect takes precedence, different patterns of change can be detected. Thus, Fig. 16(a) relates to a reaction system where there is a sudden rise in reaction rate at a specific temperature. This behaviour can occur when the ignition temperature is reached in a potentially explosive system. Fig. 16(b) illustrates the pattern of rate changes observed for reactions catalysed by enzymes. Yet other patterns of change in reaction rate with temperature can be obtained. Fig. 16(c), for example, relates to the reaction of nitric oxide (nitrogen oxide) with oxygen.

It should be noted that the reaction systems recommended for investigation in earlier sections of this book observe the type of variation shown in Fig. 15. The problem of 'accounting' for this type of variation of reaction rate with temperature remains. It is the most important pattern of change since so many different types of reaction conform to this pattern.

Historically, a number of attempts were made to account for such patterns of variation between two factors. A simple expression to account for the influence of temperature on the rate of a reaction was put forward by S. Arrhenius in 1889. It can be written in the form

$$\lg k = -Q/T + C \tag{2.22}$$

where k is a rate constant, T the temperature and Q and C are constant terms which possess specific values for a given system. The equation is more usually written in the form

$$\lg k = -E/2 \cdot 303 \, RT + \lg A \tag{2.23}$$

or in the form

$$k = A \, e^{-E/RT} \tag{2.24}$$

25

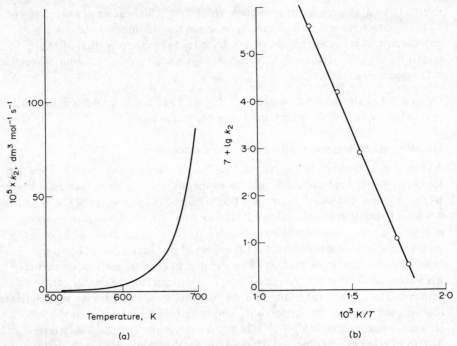

Fig. 15. Thermal decomposition of hydrogen iodide

where A is a constant, T the temperature on the Kelvin scale, R the gas constant, and E a constant with the dimensions of energy mol^{-1}. Equations (2.23) and (2.24) are forms of an equation known as the Arrhenius equation. Before considering the significance of the terms related by this equation in detail, it is instructive to use either (2.23) or (2.24) to determine the effect of temperature on the rate of

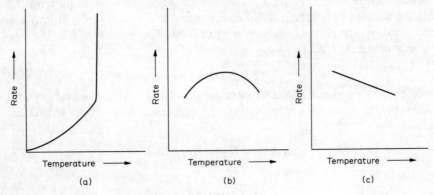

Fig. 16

reaction. Thus, if we use suffix 2 to refer to the higher temperature and the suffix 1 to refer to the lower temperature, then by using (2.24) it follows that

$$k_2/k_1 = A \ e^{-E/RT_2} / A \ e^{-E/RT_1}$$

$$= \exp\left[\frac{-E}{R}(1/T_2 - 1/T_1)\right], \tag{2.25}$$

if we assume the term A to be independent of temperature change. By rearranging equation (2.25), we can obtain a value for the term E as follows

$$E = \frac{2 \cdot 303 \, RT_2 T_1 \, \lg(k_2/k_1)}{(T_2 - T_1)}. \tag{2.26}$$

Since the gas constant, R, has the value $8 \cdot 3143 \ \text{J K}^{-1} \ \text{mol}^{-1}$, it also follows that the term E must be measured in J mol^{-1}.

PROBLEM A survey of data relating to reaction rates of a large number of reactions shows that the rate constants for a number of these systems are approximately doubled—and in some cases even trebled—when the temperature is increased by $10°C$. Apply equation (2.26) to determine the importance of the *actual* temperature interval used on the value for the constant E when (a) the reaction rate is doubled and (b) the reaction rate is trebled. It will be found convenient to compare the following ten degree temperature intervals: (i) 300 K to 310 K; (ii) 500 K to 510 K; and (iii) 1000 K to 1010 K.

The Arrhenius equation (2.23 or 2.24) can be used to account for the empirical observations on the influence of temperature on the reaction rate for many different systems. The significance of this result is somewhat difficult to interpret in terms of molecular and atomic events. However, some understanding can be gained by drawing an analogy between this variation in reaction rate with temperature and the variation of the vapour pressure of a substance with temperature. Both sets of events are summarised by equations of the same kind. In the case of the variation of vapour pressure, p, of a substance with temperature, T, the slope of the linear graph $\lg p$ vs $1/T$ is related directly to the molar heat of vaporisation of that substance: that is, to the energy that a molecule of the substance must acquire before it can drag itself away from the attraction of the molecules that surround it in the liquid phase, (cf. the discussion of this topic in Dawson, B. E. *Energy in Chemistry: an approach to thermodynamics.* Methuen, 1971. Chapter 3). By analogy, it follows that the energy term in the Arrhenius equation must correspond to the energy needed to accomplish a definite chemical process since the vapour pressure-temperature equation (the Clausius–Clapeyron equation) provides an energy term related to a definite physical process. So, equation (2.24) may be interpreted as meaning that reaction can only occur when colliding molecules, or other species, have between them a minimum critical energy E. It is noteworthy from equation (2.24) that as the value of E is increased, so the value of the exponential term, $e^{-E/RT}$, decreases: that is, the greater the energy requirement for reaction to occur, the more difficult it will be for molecules to acquire this energy. It also follows that, since the exponential term is temperature dependent, the higher the temperature the greater the exponential term becomes.

In the case of the thermal decomposition of hydrogen iodide, the Arrhenius energy of activation is $188 \cdot 7$ kJ mol^{-1} (cf. Fig. 15(b)). The Arrhenius energy of activation corresponds to the energy barrier which was seen to separate reactants and products in the elementary processes outlined in Chapter 1.

EXPERIMENT   Determine the Arrhenius energy of activation for the reaction between iodide ions and hydrogen peroxide under acid conditions. It will be found convenient to conduct experiments at $15°C$, $20°C$, $25°C$ and $30°C$, using the first of the suggested reaction mixtures described on p. 23.

Catalysis

The rate of a chemical change can be greatly altered if it is affected in any way by the addition of another substance. An accepted definition of a *catalyst* is that it is a substance which changes the rate of a chemical reaction without itself undergoing any permanent chemical change. From the experiments listed on p. 14, the addition of sulphuric acid to potassium iodide-hydrogen peroxide reaction mixtures results in an increase in reaction rate with no loss in the quantity of sulphuric acid used. (This fact can be checked fairly readily.) In the same way, the iodination of acetone is also catalysed by hydrogen ions. Indeed, it is not too difficult to use these two chemical systems to demonstrate an alternative definition of the term catalyst for reactions in solution, namely as a substance that is both a reagent and a product of a reaction.

Perhaps the simplest general interpretation of the effect of a catalyst on the rate of a chemical reaction is to employ an energy profile for the system in question and to consider the possibility of lowering the energy barrier which separates reactants from products. The concept of an energy profile for a reaction can be divorced from a purely theoretical discussion of the nature of chemical change and so it is not too difficult to envisage the possibility of alternative pathways linking reactants and products. Fig. 17 provides the simplest representation and shows that the energy of activation for the catalysed reaction is smaller than that for the non-catalysed reaction, thereby giving rise to an enhanced rate of reaction under the same set of conditions. This model compares favourably with empirical studies on a variety of systems.

Yet another possible situation is worthy of attention. The addition of a substance to a reaction mixture need not necessarily increase the rate of a reaction. The addition could result in a slowing down of the reaction rate. It might be thought that this effect must arise from the provision of a less favourable alternative reaction pathway linking reactants to products. However, this interpretation cannot account

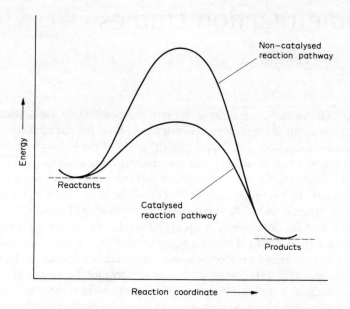

Fig. 17

for the change in reaction rate. (Why not?) The activity of 'negative catalysts' or 'inhibitors' can be accounted for in terms of the removal of either the catalytic species or the 'active' property of a material which can normally catalyse a reaction, thereby preventing reaction from taking place through a 'lower energy requirement route'. This interpretation accounts for the effect of inhibitors on the rate of a reaction. It is also of interest to consider the ways in which such a mechanism could operate in specific cases. Such matters are of great practical importance.

The subject of catalysis is a vast one. In general terms, two main types of catalysis can be distinguished: heterogeneous and homogeneous catalysis. *Heterogeneous catalysis* requires the use of an activating surface as with the catalytic oxidation of sulphur dioxide to sulphur trioxide in the contact process for the manufacture of sulphuric acid. The catalyst and the reactants exist in different phases. The catalyst in the contact process is a solid—often vanadium pentoxide or platinium—whereas the reactants and products, at the temperature and pressure under which manufacture occurs, are gases. By way of contrast, *homogeneous catalysis* occurs when both reactants and catalyst are in the same phase. Examples of this type of reaction include the iodination of acetone in the presence of acid (cf. p. 19) and the oxidation of sulphur dioxide by the oxides of nitrogen in the Chamber process for the manufacture of sulphuric acid.

Additional information on catalysis is provided in another volume in this series of monographs—Robertson, A. J. B. *Catalysis in Chemistry*. Methuen, 1972.

3 Some reaction studies

The investigation of the mechanism of a reaction depends upon a systematic use of all known facts about the system under review: namely, the consumption of the reactants under known conditions; the identification of the products of reaction and the proportion in which these are produced under specific conditions; the stoichiometry of the reaction; and, the factors which influence the rate of reaction. In the previous chapter, we have seen that the latter include the concentrations of the reactants; the temperature of the system; and the influence of catalysts. In addition, we have seen that the influence of temperature on the reaction rate can be accounted for in many instances by the Arrhenius equation and this leads to a measure of the additional energy required to effect reaction, the energy of activation. The descriptive accounts of reaction studies which follow also require the use of other forms of enquiry which do not necessarily depend upon the measurement of reaction rate. However, it should be noted that matters which may be most conveniently considered as purely theoretical approaches to the study of reaction rates are considered separately in Chapter 4.

Reaction mechanism

The term 'reaction mechanism' is used to convey any detail about the progress of a reaction from reactants to products which is not expressed by the overall stoichiometric equation for that change. Some of the details are obvious. For example, an intermediate may be formed and it may even be possible to isolate it and subsequently to use it to form the final products of the 'normal' reaction. In most instances where reactions proceed through a number of 'steps' or 'stages', the actual isolation of intermediates is rarely feasible although their presence can be detected through the use of physical (often spectroscopic) methods of analysis. In addition, it can be shown that intermediates are transient species and exist only in extremely low concentration. A convenient demonstration of the formation and decomposition of intermediates may be provided by the methylene blue oxidation of glucose under alkaline conditions in a sealed vessel, (the so-called 'blue bottle' experiment—see Campbell, J. A. *J. Chem. Educ.*, 1963. **40**, 578).

In Chapter 1, it was shown that an energy profile may be used to describe the course of a reaction and simple forms of profile appear in Fig. 6 and in Fig. 17. In these cases, the highest point on the energy profile—leading from reactants to products—is defined as the transition state, a species which cannot be isolated or even detected by analytical techniques.

Fig. 18 relates the terms intermediate and transition state to an energy profile for a hypothetical reaction. The reaction possesses only one intermediate and, in consequence, must have two transition states. From Fig. 18 it can be seen that an intermediate is defined as a minimum on the energy profile of the reaction whereas a transition state corresponds to a maximum. In this instance, the two transition states are shown to exist with very different energy requirements. The transition

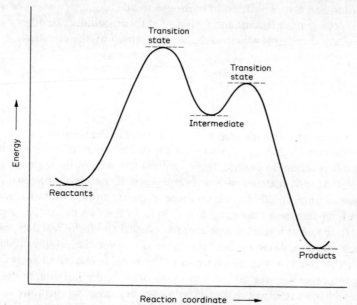

Fig. 18. The transition state of the slow step is always the highest point on the energy profile leading from reactants to products

state of the slow step of the mechanism will be the highest point on the energy profile *if* the Arrhenius A factor is identical for both steps.

As will become apparent when we refer to actual reaction studies these ideas do not preclude a close structural resemblance between a reaction intermediate and a transition state. Indeed, the similarity between an intermediate and a transition state will be most marked when only a small energy difference exists between the two species. In other words, the intermediate will be highly unstable when the energy valley in which the species exists is shallow. A wide experience of chemical reactions reveals that the relative stability of a reaction intermediate (which we could define as the energy difference between the intermediate valley on the reaction profile and the lowest transition state—cf. Fig. 18) can vary over quite a wide range.

These general ideas will now be used to interpret a number of separate studies of specific reaction mechanisms.

The hydrogen, iodine, hydrogen iodide system

The synthesis of hydrogen iodide from its constituent elements and its decomposition provide useful illustrations of the general ideas which have been discussed so far. The system has the added advantage in that it indicates the need to re-assess accepted scientific interpretation periodically by making use of more recent advances in technique and ideas. Until relatively recently, it was considered that both the formation and decomposition of hydrogen iodide took place when two reactant molecules, possessing sufficient energy between them, collided. In other words, that reaction occurred in terms which could be summarised by the equations

$$H_2 + I_2 = 2HI$$

and

$$2HI = H_2 + I_2$$

Now a greater insight into the reaction mechanism has been obtained.

During the last 75 years, a number of chemists have investigated the kinetics of this reversible reaction. In general, these studies have been of the type needed to establish the order of reaction with respect to each of the species present. Reaction vessels used for such studies have been made of quartz and it is not unusual to operate at temperatures in the range 500–800 K. In the case of the decomposition of hydrogen iodide at a given temperature, it is usual to fill the reaction vessel to an appropriate pressure. After a convenient interval, the vessel is rapidly chilled to prevent further reaction and the contents of the vessel subjected to analysis. By repeating this procedure for different intervals of time, the variation in the concentrations of the participating species with time at a given temperature can be determined. The results of experiments show that, at low conversions, the rate of reaction (considered as the rate of formation of hydrogen) is proportional to $[HI]^2$.

The reverse reaction presents some practical problems in view of the need to use iodine vapour at relatively high pressures. The procedure requires that the reactants, hydrogen and iodine vapour, are heated in a sealed quartz vessel at a known temperature for a known time and then chilled rapidly so as to permit the contents of the vessel to be determined at that instant. As in the former reaction study, a detailed result can only be obtained from a series of separate experiments. It is found that for this reaction, the rate of reaction—considered as the rate of disappearance of hydrogen—is proportional to the product $[H_2][I_2]$.

Both sets of experimental results suggest that the net rate of formation of hydrogen from a mixture of all three gases not at equilibrium will be given by

$$d[H_2]/dt = k_2[HI]^2 - k_3[H_2][I_2] \tag{3.1}$$

at a known temperature. It is noteworthy that (3.1) can be used to account for the two sets of actual results. (How?) The rate constant k_2 refers to the decomposition of hydrogen iodide and the rate constant k_3 refers to the synthesis of hydrogen iodide. Now as each reaction mixture at a given temperature reaches equilibrium the concentrations of each of the species present achieve special values, $[H_2]_e$, $[I_2]_e$ and $[HI]_e$, such that at equilibrium

$$k_3[H_2]_e[I_2]_e = k_2[HI]_e^2.$$

It follows that

$$K_c = \frac{k_3}{k_2} = \frac{[HI]_e^2}{[H_2]_e[I_2]_e} \qquad (3.2)$$

where K_c is the equilibrium constant at a given temperature expressed in terms of the concentrations of the various species. The values of the rate constants k_2 and k_3 can be expressed in the manner derived from the Arrhenius equation (2.24): thus,

$$k_3 = 5 \cdot 0 \times 10^{11} \times \exp(-167\,400/RT) \text{ dm}^3 \text{ mol}^{-1} \text{ s}^{-1} \qquad (3.3)$$

$$k_2 = 9 \cdot 0 \times 10^{10} \times \exp(-184\,000/RT) \text{ dm}^3 \text{ mol}^{-1} \text{ s}^{-1} \qquad (3.4)$$

It follows that the synthesis of hydrogen iodide is an exothermic reaction (cf. Fig. 6a), such that the net energy change between reactants and products is 16 600 J, and that the mechanism of the reaction would appear to be similar to one of those portrayed in Fig. 1. From purely energetic considerations, it is apparent that the mechanism cannot involve the dissociation of both hydrogen and iodine into atoms prior to the formation of hydrogen iodide molecules, (cf. Fig. 19). However, it is worth noting that the energy of activation for the formation of hydrogen iodide is greater than the bond dissociation energy of iodine.

The classical work on this system was done by M. Bodenstein (1894-9) and most of this work was conducted in the temperature range 250°C to about 500°C. Indeed, most of the experimental data collected over some 75 years is consistent with the simple bimolecular mechanism which has been described, although at least one significant discrepancy in the experimental rate constants was pointed out in 1930 for the thermally activated reaction. A careful review of data shows that the Arrhenius equation does not yield a linear relationship between the logarithm of the rate constant and the reciprocal of the temperature of the reaction. A slightly curved plot is obtained and this implies that the activation energy changes with temperature. An explanation of this effect was put forward by S. W. Benson and R. Srinivasan in 1955 and subsequently confirmed experimentally by J. H. Sullivan in 1959 when it was shown that the effect was consistent with two concurrent mechanisms in the temperature range 600 to 700 K. These were considered to be

Mechanism A:

$$(1) \quad H_2 + I_2 \xrightarrow{k_1} 2HI$$

$$(2) \quad H_2 + 2I \xrightarrow{k_2} 2HI$$

Mechanism B:

$$I + H_2 \rightarrow HI + H$$

$$H + I_2 \rightarrow HI + I$$

$$H + HI \rightarrow H_2 + I$$

$$I + HI \rightarrow H + I_2$$

Although discussions on Mechanism A recognised that reaction (2) might occur, it was not until 1967 that calculations based on reaction (2) achieved any real effect. A little thought will show that in a thermally activated system, reactions (1) and (2) are kinetically indistinguishable since iodine atoms must clearly be in thermal equilibrium with iodine molecules. This can be expressed as follows—

$$[I] = K_D [I_2]^{\frac{1}{2}}$$

where K_D is the equilibrium constant for the dissociation. It follows that

$$\text{Rate of formation of HI} = d[HI]/dt = k_2 [I]^2 [H_2]$$

$$= k_2 K_D^2 [I_2] [H_2]$$

$$= k [I_2] [H_2],$$

where k_2 is defined by mechanism A above and k replaces $k_2 K_D^2$. It was shown that mechanism A is dominant below about 800 K and that mechanism B could account for the slight curvature in the Arrhenius plots.

In 1967, J. H. Sullivan argued that if reaction (2) in mechanism A occurred in thermally induced reactions, then its rate could be measured by using photochemically produced iodine atoms at a temperature where thermally induced reactions do not occur. Accordingly, he irradiated hydrogen–iodine mixtures with light of wavelength 578 nm which was known to produce iodine atoms from iodine. His experiments showed that the concentration of hydrogen iodide produced by this means was proportional to the square of the iodine atom concentration. He then went on to calculate rate constants for reactions carried out at 417·9 K, 480·7 K and 520·1 K, basing his calculation on the termolecular reaction (2) above. Sullivan also treated experimental data for thermal systems in the range 633 to 738 K on the same hypothesis. The rate constants obtained gave the same Arrhenius parameters as the rate constants from the photochemical experiments. All of this suggests that reaction (1) of mechanism A does not occur. Maybe at some time, further work will reveal a factor which will enable us to rule out one, or even both, of the other reaction pathways mentioned. Until such information becomes available, we are justified only in claiming that experimental data are consistent with such mechanisms.

Fig. 19 summarises the results of experiments up to 1967 on an appropriate energy profile. Fig. 20 reproduces Sullivan's critical Arrhenius plot for the reaction system.

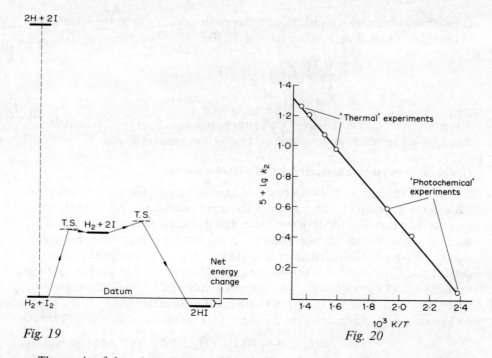

Fig. 19   Fig. 20

   The result of these investigations does not alter our view on the composition of
the transition state for the slow step of these two reactions: it is still H_2I_2. (This
follows directly from the kinetic order of the reactions.) The new work adds a
further dimension to our understanding of the transition state for these reactions.
We know that the transition state for the slow step could be either a linear structure
such as

$$I \ldots H \ldots H \ldots I$$

or a staggered structure which need not have all four atoms coplanar, such as

I
.
.
.
H . . . H
.
.
.
I

35

rather than a planar trapezoidal structure dependent entirely upon the actual sizes of the iodine and hydrogen atoms, (cf. Fig. 1), like

$$
\begin{array}{c}
\text{I} \ldots \ldots \text{I} \\
\text{H} \ldots \text{H}
\end{array}
$$

The precise nature of the complex has yet to be resolved. Even so, these model structures are helpful since they lead to ideas on how molecules react.

Gas phase pyrolysis of chloroethane and related reactions

This second example has been selected to extend an appreciation of the ideas already discussed in this chapter. The example is also representative of a large number of gas phase investigations on the haloalkanes. The significance of work done in the gas phase is not difficult to understand for, under suitable conditions, it becomes possible to investigate the behaviour of single molecules, uninfluenced by other parts of the reaction system. Further, the effect of substitution in a parent molecule upon the rate of a given reaction can be studied without the potential complications which might arise from the co-operative effects of a solvent.

The stoichiometric equation for the pyrolytic decomposition of chloroethane is

$$CH_3 . CH_2Cl = CH_2:CH_2 + HCl$$

and is an example of the general type of reaction

$$
\underset{R^2}{\overset{R^1}{>}}CH-\underset{\underset{X}{|}}{C}\underset{R^3}{\overset{R^4}{<}} \longrightarrow \underset{R^2}{\overset{R^1}{>}}C=C\underset{R^5}{\overset{R^4}{<}} + HX
$$

where X can be either Cl, Br, I, HCO_2, CH_3CO_2, $ClCO_2$ and the various Rs are capable of a wide range of variation. The general stoichiometry for such reactions conforms to the above equation except in the case of iodo-compounds,

$$2R^1R^2CH . CR^3R^4I \rightarrow R^1R^2CH . CHR^3R^4 + R^1R^2C:CR^3R^4 + I_2$$

In the case of chloroethane, it is not too difficult to see from a model of the molecule that the rotation of the carbon–carbon bond could play an important role in the reaction mechanism: it would permit a special alignment of bonds in the transition state is this consisted of a species of the same composition as the molecule chloroethane. Experiments designed to assess the order of reaction for this system reveal that the reaction is indeed a first order reaction. Accordingly, the transition state for this reaction must be the result of a redistribution of energy within the chloroethane molecule and, given that the Arrhenius energy of activation for this reaction measured over the temperature range 470–520 K is about 237 kJ mol^{-1}, we could express the mechanism of the reaction as follows

$$CH_3 . CH_2Cl \xrightarrow{E = 237 \text{ kJ}} \left[\begin{array}{c} H \diagdown \quad \diagup H \\ \quad C = C \\ H \diagup \mid \quad \mid \diagdown H \\ \quad H \cdots Cl \end{array} \right] \longrightarrow \begin{array}{c} H \diagdown \quad \diagup H \\ C = C \\ H \diagup \quad \diagdown H \end{array} + HCl$$

transition
state

The redistribution of energy within the molecular species prior to decomposition can be interpreted in a number of ways. Modern thought on this matter suggests that the redistribution leads to a concentration of energy in the carbon–halogen bond. Given this hypothesis, the carbon–halogen bond must be extended and accompanied by a polarisation in the sense $C^{\delta+} X^{\delta-}$. For the general case, it follows that the transition state could be described by one or more of the following structures

$$\begin{array}{cccc} \diagdown C - C \diagup & \diagdown C - \overset{+}{C} \diagup & \diagdown C = C \diagup & \diagdown C = C \diagup \\ \mid \quad \mid & \mid \quad \mid & & \\ H \quad X & H \quad X^- & H^+ \quad X^- & H - X \\ (I) & (II) & (III) & (IV) \end{array}$$

The *actual* structure must depend on the contributions made by the various possibilities: thus, if (*I*) and (*IV*) make the greatest contribution, then (*V*) would best represent the transition state; if (*II*) and (*III*) make the greatest contribution, then (*VI*) would serve as the best representation.

$$\begin{array}{cc} \diagdown C = C \diagup & \left(\diagdown C = C \diagup \right)^+ \\ H \cdots X & \quad \quad H \quad \diagup X^- \\ (V) & (VI) \end{array}$$

Clearly, such representations are more complex than the simple homolytic fission of a covalent bond, such as

$$A:B \rightarrow A \cdot + \cdot B; \text{ energy requirement } D(A-B)$$

or heterolytic fission, such as

$$A:B \rightarrow A^+ + :B^-; \text{ energy requirement } D(A^+ B^-)$$

However, by comparison, it will be noted that (*V*) is essentially a homolytic species whereas (*VI*) is essentially heterolytic. Structure (*V*) owes its reactivity as a free radical to a deficiency of electrons whereas (*VI*) owes its reactivity as an ionic species to the same cause.

Recent work by A. Maccoll (1959 and 1965) discussed the transition state for these gas phase elimination reactions in terms of the Arrhenius energies of activation for elimination, $E(H-X)$, and the heterolytic bond dissociation energies, $D(R^+ X^-)$, for a series of haloalkanes. The linear relationship which obtains for these two parameters in a number of instances (including chloroethane) indicates some support on energetic grounds for transition states like (*VI*) (cf. Fig. 21). However, as can be seen from Fig. 22, a linear relationship is also observed for homolytic bond dissociation energies, $D(R-X)$, for three sets of halides, each set being a related α-methylated series of haloalkanes. These findings, together with those from many other different types of study, suggest that the structures for the various transition states approximate to a blend of features shown by the representations (*V*) and (*VI*), including the significant contribution through ionisation in the gas phase though not of ionic dissociation. It will also be noted (from both Fig. 21 and Fig. 22) that the rate of elimination must be determined primarily by the environment of the carbon–halogen bond.

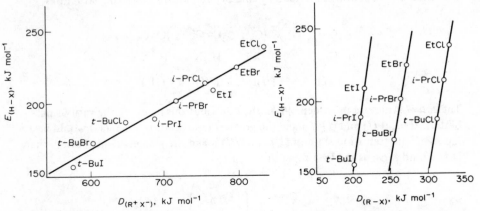

Fig. 21. Activation energies for eliminations, E(H–X), plotted against the heterolytic bond dissociation energies, D(R⁺ X⁻), (after A. Maccoll, 1966)

Fig. 22. Activation energies for eliminations, E(H–X), plotted against the homolytic bond dissociation energies, D(R–X), (after A. Maccoll, 1966)

EXPERIMENT   It will be appreciated from the two reaction studies described so far that experimental techniques for gas phase studies are difficult and require a great deal of time. This need not be the case as has been shown for the thermal decomposition of di-*t*-butyl peroxide (cf. suggestions for simplified procedures by Price, A. H. and Baker, R. T. K., *J. Chem. Educ.*, 1965, **42**, 614; Guillory, W. A., *J. Chem. Educ.*, 1967, **44**, 514; Trotman-Dickenson, A. F., *J. Chem. Educ.*, 1969, **46**, 396; Ellison, H. R., *J. Chem. Educ.*, 1971, **48**, 205). The rate determining step for this decomposition is a true unimolecular homogeneous reaction and the activation energy for this process has a simple physical significance: it is equal to the strength of the 0–0 bond.

So far, no account has been given of the interpretation of gas phase reactions which observe first order kinetics using a collision mechanism. A little thought will enable the reader to appreciate that the rate of such a reaction cannot be proportional to the collision frequency since the rate is proportional to the first and not the second power of the concentration of the reactant. (An elementary account of the theory of the rates of unimolecular reactions is given in Chapter 4.)

Hydrolysis of 2-bromo-2-methylpropane and related reactions

It is appropriate to consider some of the other reactions of the haloalkanes in addition to the pyrolytic decomposition reactions outlined in the previous section. The reaction of 2-bromo-2-methylpropane with solvent can lead to a greater understanding of the mechanisms of organic reactions in solution. It also provides an example of the successful use of kinetic orders of reaction as a criterion of reaction mechanism.

EXPERIMENT   (1) Investigate the solvolysis of 2-bromo-2-methylpropane in aqueous ethanol and isolate samples of the various products of reaction. Attempt the investigation at different temperatures and with different mixtures of ethanol and water as solvent.

(2) Compare the reactions of various haloalkanes with silver ions in solution. It will be found convenient to use comparable concentrations of haloalkanes in aqueous ethanol at about 60°C and to add 0·1M silver nitrate solution (also at 60°C) to each reaction mixture. A variety of simple test tube reactions can be devised so as to compare (a) the effect of structure of the alkyl group and (b) the effect of changing the halogen on the rate of reaction.

(3) The results of (2) above suggest that ions may be formed in many instances. Examine the possibility of using a conductance bridge and cell to follow the hydrolysis of either 2-bromo-2-methylpropane or 2-chloro-2-methylpropane in aqueous ethanol solution. What other methods of analysis could be adapted to follow the course of this hydrolysis?

Many different preliminary experiments of the kind indicated above provide much valuable information to supplement formal determinations of kinetic order of reaction and of the energy of activation. Thus, it is not too difficult to show that in the case of the solvolysis of 2-bromo-2-methylpropane in aqueous ethanol, three carbon compounds are formed: an alkene, an ether and an alcohol. The rate of solvolysis can also be shown to depend on the composition of the solvent, the rate of reaction increasing with a decrease in the proportion of ethanol used in the solvent. However, rate studies leading to the determination of the order of reaction are most conveniently carried out with dilute solutions and these make interpretation of experimental results somewhat more difficult. It is found that the reaction is first order with respect to 2-bromo-2-methylpropane.

These results can be accounted for in terms of a reaction mechanism which possesses a slow step and subsequent fast steps. (The effect of such a mechanism on the observed features of a reaction can be likened to the effect of allowing water

39

to flow through a series of tubes, if one of the tubes possesses a much smaller diameter exit hole than the others. Such tubes can be constructed from lengths of plastic tubing, of about 1″ diameter, with one end open and the other closed save for a central hole of either 0·25″ or 0·125″ diameter.) Thus, mechanism A may be used to account for all of the observed facts during the solvolysis of 2-bromo-2-methylpropane

Mechanism A:

$$(CH_3)_3C \cdot Br \longrightarrow (CH_3)_3C^+ + Br^- \text{ (slow)}$$

and

$$(CH_3)_3C^+ + HOH \longrightarrow (CH_3)_3C \cdot OH + H^+ \text{ (fast)}$$

or

$$(CH_3)_3C^+ + C_2H_5OH \longrightarrow (CH_3)_3C \cdot O \cdot C_2H_5 + H^+ \text{ (fast)}$$

or

$$H\overset{\frown}{-}CH_2 \cdot C^+(CH_3)_2 \longrightarrow H^+ + H_2C=C(CH_3)_2 \text{ (fast)}$$

This mechanism assumes that the slow step is common to all subsequent steps whether these lead to the formation of the products of substitution or elimination reactions.

Mechanism B may also be possible under the conditions used to determine the order of reaction. This is a bimolecular mechanism and takes place in a single step. Since the solvent is in great excess during the determination of the order of reaction, the overall concentration of solvent will remain substantially constant during the determination, thereby allowing the system to exhibit first order kinetics.

Mechanism B:

$$HOH + (CH_3)_3C \cdot Br \rightarrow (CH_3)_3C \cdot OH + HBr \text{ (slow)}$$

or

$$C_2H_5OH + (CH_3)_3C \cdot Br \rightarrow (CH_3)_3C \cdot O \cdot C_2H_5 + HBr \text{ (slow)}$$

or

$$Y + (CH_3)_3C \cdot Br \rightarrow YH^+ + H_2C=C(CH_3)_2 + Br^- \text{ (slow)}$$

where Y is a nucleophilic reagent (e.g. OH^-).

The study of the kinetics of a reaction usually involves the determination of the kinetic order with respect to each reactant. By this means it is hoped to determine the molecularity of the slow step and thereby determine the composition of its transition state. This procedure can fail if a change in the concentration of a reactant occurs by virtue of changes in the medium. It follows that we need some test to distinguish between the two types of mechanism, A and B. Evidence to support mechanism A can be obtained from a number of empirical tests.

If, for example, the results of rate studies were available for both the hydrolysis of 2-bromo-2-methylpropane and of 2-chloro-2-methylpropane, then − provided

40

mechanism A applies — the same carbonium ion will be produced and will undergo substitution and elimination reactions. The rate constant for the *total* unimolecular reaction, k_1, must equal that for the substitution and elimination reactions:

$$k_1 = k(S_N1) + k(E1) \qquad (3.5)$$

Clearly, the value of k_1 can vary and will depend on whether the chloro-compound or the bromo-compound is used. However, since a unimolecular mechanism (mechanism A) has been assumed, the proportion $k(E1)/k_1$ (in which the formed carbonium ion will decompose so as to complete the elimination process) should be almost independent of the halogen or other group present in the reactant. This hypothesis can be checked against experimental data. If it is true, then the non-carbonium ion forming group X (where X can be Cl, Br, I, etc.) must pass beyond simple separation when the fate of the transition state of heterolysis is decided. The group X need not be very far away from the carbonium ion and could influence the relative rates of the alternative reactions of the carbonium ion. However, this could only lead to minor variations in the ratio $k(E1)/k_1$ whereas by changing X there are changes in order of magnitude for k_1, $k(E1)$ and $k(S_N1)$. Table 2 lists data to show the comparative constancy of the proportion of alkene formed, $k(E1)/k_1$, despite the large changes in overall rate.

Other evidence supporting the unimolecular nature of the solvolysis of 2-bromo-2-methylpropane and of 2-chloro-2-methylpropane could be obtained from experiments in which the haloalkane was in such large excess that its concentration did not change appreciably during the reaction. The results of such experiments show that the rate of reaction in alkaline solution is constant and independent of the concentration of alkali used. Furthermore, when the alkali present is neutralised by the halogen acid formed during the reaction, the solution acquires acidity. Under these circumstances, it is found that the reaction rate continues unaffected by this change.

Another source of evidence for the unimolecular mechanism came from experiments in a solvent which had an ionising power towards the haloalkanes as great or greater than that of water. The addition of water in small quantities could therefore have no profound change in the ionising power of the medium towards these compounds. By using formic acid as solvent, L. C. Bateman and E. D. Hughes (1937) were able to show that the rate of solvolysis of 2-chloro-2-methylpropane in moist formic acid was independent of small amounts of added water. Their results have been summarised in Fig. 23 which shows the concentration of chloride ions formed during the reaction as a function of time. It can be seen that the initial rate of reaction is virtually the same irrespective of the water content of the solvent. The subsequent curvature in each graph is interpreted in terms of the reversibility of the reaction. It was also found that as the percentage of water present in the solvent was increased, so the percentage of alkene formed decreased. Thus, when the solvent had a high water content, it was found that the only product formed from the alkyl group was 2-methylpropan-2-ol, $(CH_3)_3C \cdot OH$. We might therefore safely conclude

Table 2 Proportion of alkene formed in unimolecular reactions of haloalkanes (after C. K. Ingold, 1953)

Compound	solvent	temperature °C	$10^5 k_1$ s^{-1}	$k(E1)/k_1$
$(CH_3)_3C.Cl$	60% aq-alc	25·0	0·854	0·168
$(CH_3)_3C.Br$	60%	25·0	37·2	0·126
$(CH_3)_3C.I$	60%	25·0	90·1	0·129
$\begin{array}{c} CH_3 \\ \cdot \\ H_3C.C.Cl \\ \cdot \\ CH_2.CH_3 \end{array}$	80% aq-alc	25·2	1·50	0·333
$\begin{array}{c} CH_3 \\ \cdot \\ H_3C.C.Br \\ \cdot \\ CH_2.CH_3 \end{array}$	80%	25·2	58·3	0·262
$\begin{array}{c} CH_3 \\ \cdot \\ H_3C.C.I \\ \cdot \\ CH_2.CH_3 \end{array}$	80%	25·2	174	0·260

that the same rate determining step applies to all of the reaction mixtures studied and is, in fact, the ionisation step of mechanism A.

It remains for us to consider the influence of solvent on the slow step of mechanism A. The dissociation of an individual molecule of 2-bromo-2-methyl-propane will occur only when it possesses sufficient energy as the result of molecular collision and interaction with solvent molecules. If this acquired energy is used to raise the vibrational level of the carbon–bromine bond, ionic dissociation is feasible. The influence of polar solvent molecules on the incipient bromide ion present in the reactant can be visualised in terms of electrostatic forces or of hydrogen bonding, as may be appropriate. Accordingly, although the slow step of mechanism A does not specify a use of solvent molecules, a change in solvent will alter the energies of the reactant molecules and of the transition state for that step. These changes will depend on differences in solvation between the reactant and this first transition state. Solvolysis experiments in mixed solvents (cf. p. 39) reveal that the rate of solvolysis of 2-bromo-2-methylpropane changes with the composition of the solvent, the rate increasing with the increase in polarity of the mixture. For water–ethanol mixtures, the rate increases with an increase in the proportion of water used and

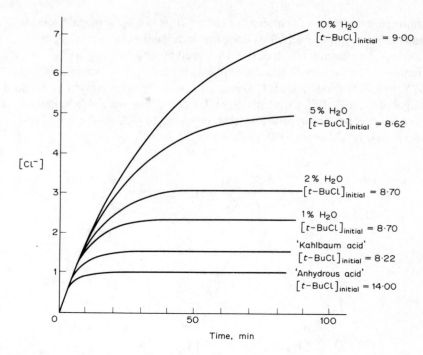

10 % H₂O
$[t-\text{BuCl}]_{\text{initial}} = 9\cdot00$

5 % H₂O
$[t-\text{BuCl}]_{\text{initial}} = 8\cdot62$

2 % H₂O
$[t-\text{BuCl}]_{\text{initial}} = 8\cdot70$

1 % H₂O
$[t-\text{BuCl}]_{\text{initial}} = 8\cdot70$

'Kahlbaum acid'
$[t-\text{BuCl}]_{\text{initial}} = 8\cdot22$

'Anhydrous acid'
$[t-\text{BuCl}]_{\text{initial}} = 14\cdot00$

$[\text{Cl}^-]$

Time, min

Fig. 23. Rate of formation of chloride ions from 2-chloro-2-methylpropane in formic acid solutions at 15°C (after L. C. Bateman and E. D. Hughes, 1937)

(Note: all concentrations are expressed in the same units — cm^3 $0\cdot1 - NH_4CNS$ per 5 cm^3 solution.)

this can be interpreted in terms of an increase in solvation effects leading to a relative lowering of the energy of the transition state for the first step of mechanism A.

For other reactions of the type considered in this section, the process of ionisation for a unimolecular mechanism is also an endothermic step. Subsequent interaction with solvent molecules can, in appropriate instances, be more complex than indicated by the example discussed above. For example, a reactant R—X could lead to the formation of an intermediate with the properties of an ion-pair. Then, the process of solvation could lead to the gradual separation of ionic species. Thus, an intimate ion-pair (*I*) could lead to the formation of a solvent separated ion-pair (*II*) before going on to the formation of dissociated ionic species (*III*) as indicated in mechanism C and in Fig. 34.

Mechanism C:

$$R–X \rightleftharpoons R^+ X^- \rightleftharpoons R^+ \| X^- \rightleftharpoons R^+ + X^-$$

$$(I) \qquad (II) \qquad (III)$$

43

This representation, in the interest of clarity, does not show specific solvent molecules. Even so, it is much more complex than the first step of mechanism A. It is also no mere speculation since it is supported by evidence acquired from a study of a number of systems by S. Winstein and his co-workers. For example, in 1951 W. G. Young, S. Winstein and H. L. Goering investigated the acetolysis of 3-chloro-3-methylbut-1-ene (1,1-dimethylallyl chloride), (*IV*). They were able to demonstrate that the reaction was accompanied by rearrangement to 4-chloro-2-methylbut-2-ene (3,3-dimethylallyl chloride), (*VI*), as shown in the scheme

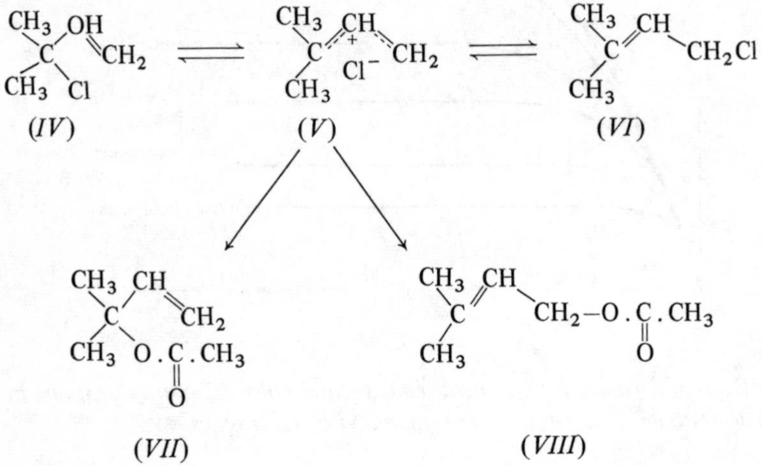

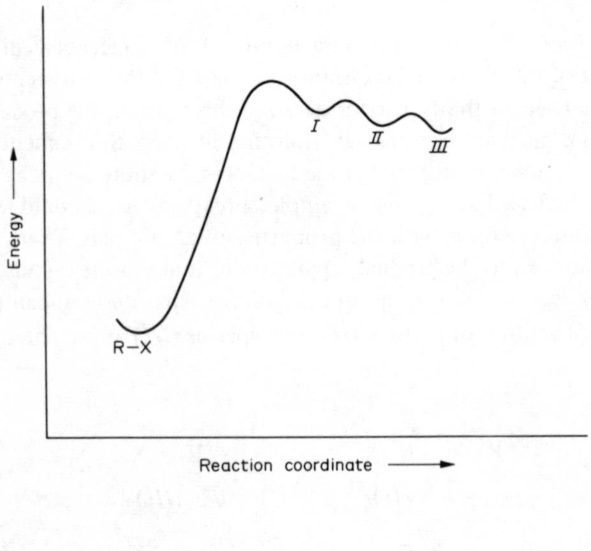

(*IV*) (*V*) (*VI*)

(*VII*) (*VIII*)

Fig. 24

The addition of chloride ions was found *not* to influence the formation of (*VI*), (*VII*) or (*VIII*), thereby indicating that the reaction does not proceed through a dissociated carbonium ion but rather through an ion-pair, (*V*). Since the rate of formation of (*VI*) is greater than the rate of formation of either (*VII*) or (*VIII*), it is suggested that the process is one requiring a 'recombination' of ions rather than of solvolysis. When ethanol was used as solvent, it was found that the isomerisation did not occur. This result should not be unexpected since ethanol has a greater power to donate an electron pair to carbonium ion (*V*) than has acetic acid.

All of this information suggests that the significance of mechanism C for a chemist examining this group of reactions is that he must question the origin of the final products from postulated reaction intermediates. In the case of mechanism C, products of reaction could arise from three such intermediates rather than one as with mechanism A. Indeed, given particular circumstances, each intermediate could possess specific stereochemical properties which would be reflected in the products of a reaction. Precise details of such reaction systems need not concern us: we need only sense the potential of this mechanistic situation.

Our revised general interpretation for unimolecular substitution reactions finds support from yet other evidence. For example, the solvolysis of 1-bromobicyclo-[2,2,2]octane, (*IX*), cannot conceivably require the formation of a carbonium ion specifically solvated at the rear of the C—Br bond.

(*IX*)

This species is comparatively unreactive due, it is felt, to the difficulty of forming a carbonium ion at a bridgehead structure of this type. The solvolysis must proceed through the formation of an ion-pair, as suggested in mechanism C.

It is noteworthy that not all haloalkanes undergo solvolysis in this way. Indeed many such compounds observe mechanism B, a process requiring but a single bimolecular step — as with the hydrolysis of bromomethane. The use of optically active species and of radioactive isotopes provides some interesting stereochemical features for mechanism B. The mechanism has been reported in great detail, particularly by E. D. Hughes, C. K. Ingold and their co-workers in the period immediately before and after the second World War. In essentials, this bimolecular mechanism for substitution reactions requires the formation of a transition state such as either (*X*) or (*XI*). It will be appreciated that (*X*) requires an inversion of stereochemical form whereas (*XI*) requires a retention of stereochemical form

45

during the reaction. The symbols X and Y used in both (*X*) and (*XI*) refer to ingoing and outgoing groups respectively.

$$\underset{(X)}{\overset{\overset{R^2\quad R^3}{\diagup}}{X\ldots C\ldots Y}}\underset{R^1}{|}\qquad\qquad\underset{(XI)}{\overset{R^1\quad X}{\underset{R^2\diagup\,\diagdown R^3 Y}{C}}}$$

Perhaps the simplest demonstration of (*X*) was obtained by Hughes and his co-workers in the period 1935–38 by an isotopic tracer method for the halogen exchange reaction between halide ions and haloalkane molecules. If an asterisk is used to indicate radioactivity, then the racemisation of 2-iodooctane by sodium iodide in dry acetone can be expressed by

$$*I^- + CH_3.CH_2.CH_2.CH_2.CH_2.CH_2.\overset{\overset{H}{\cdot}}{\underset{\underset{I}{\cdot}}{C}}.CH_2 \rightarrow$$

$$I^- + CH_3.CH_2.CH_2.CH_2.CH_2.CH_2.\overset{\overset{H}{\cdot}}{\underset{\underset{*I}{\cdot}}{C}}.CH_3$$

The racemisation of 1-bromo-1-phenyl-ethane by lithium bromide in dry acetone provides another example,

$$*Br^- + \langle\text{C}_6\text{H}_5\rangle\!-\!\overset{\overset{H}{\cdot}}{\underset{\underset{Br}{\cdot}}{C}}.CH_3 \longrightarrow Br^- + \langle\text{C}_6\text{H}_5\rangle\!-\!\overset{\overset{H}{\cdot}}{\underset{\underset{*Br}{\cdot}}{C}}.CH_3$$

A further example is the racemisation of 1-bromopropanoic acid by lithium bromide in dry acetone.

$$*Br^- + \underset{H\quad Br}{\overset{HO_2C\quad CH_3}{C}} \longrightarrow Br^- + \underset{H\quad *Br}{\overset{HO_2C\quad CH_3}{C}}$$

In each instance, it was found that

(1) substitution occurs with second order kinetics and so takes place by a bimolecular mechanism;

(2) the rate of loss of optical activity requires that *every* individual act of bimolecular substitution invert configuration and so pass through a transition state of the type shown by (*X*).

46

It is of interest to note that for the examples considered, the split bond $X \ldots C \ldots Y$ holding the incoming and outgoing groups will have a planar surface of zero electronic density in which all three C–R bonds can lie. In those instances where X is not the same species as Y, the surface would not be quite planar. The energy profile for reactions observing mechanism B is of the type shown by Fig. 6.

In conclusion, it is also worth recording that extensive studies of the type reported in this section led to a systematic nomenclature for the various types of mechanism observed. Thus, in its simplest interpretation, mechanism A (cf. p. 40) may be equated with two separate mechanisms both of which depend on a common unimolecular slow step: the nucleophilic substitution mechanism referred to as $S_N 1$ and an elimination mechanism known as E1 (cf. p. 41). Mechanism B (cf. p. 40) also covers two separate types of reaction mechanism: a bimolecular nucleophilic substitution, $S_N 2$, and a bimolecular elimination, E2. Other types of mechanism are possible and are described in a number of advanced monographs and general textbooks.

Hydrolysis of simple esters under alkaline conditions

The hydrolysis of a simple ester under acid conditions formed the basis of one group of experiments in Chapter 2. These experiments were conducted to illustrate methods of obtaining orders of reaction. The purpose of this section is to extend an appreciation of this type of reaction and to indicate the means whereby isotopic analysis was used to supplement kinetic measurements in deriving the mechanism of the hydrolysis under alkaline conditions.

EXPERIMENT   Compare the rates of hydrolysis of methyl formate (methyl methanoate) and of methyl acetate (methyl ethanoate) under alkaline conditions. This may be conveniently carried out in a stoppered flask by adding 0·01 mole of the ester (i.e. 0·61 cm^3 methyl formate *or* 0·79 cm^3 methyl acetate) to a mixture of 0·005 mole sodium hydroxide in water (i.e. 50 cm^3 0·10M NaOH) and a few drops of 1% phenolphthalein indicator. The time required to decolorise the indicator in each instance should be recorded. Can you account for the markedly different times of reaction? Is it necessary to measure out the volumes of ester used with the degree of precision suggested? (If not, why not?) Can you devise a mechanism to account for your observations?

Ester hydrolysis has been studied extensively over a long period. However, it was not until 1934 when M. Polanyi and A. L. Szabo used water enriched in ^{18}O, a stable isotope, that it could be shown that oxygen from the reaction medium did not appear in the alcohol formed during the reaction. The reaction used for this investigation was the hydrolysis of pentyl acetate (n-amyl acetate) under alkaline conditions.

$$CH_3 . CO . OC_5H_{11} + H{-}^{18}O{-}H \xrightarrow{\text{OH}^-} CH_3 . CO . {}^{18}OH + HO . C_5H_{11}$$

Subsequently, M. Cohn and H. C. Urey reported that this ester gave no exchange with water enriched with ^{18}O. Somewhat later (in 1956), doubt was expressed on the validity of the evidence provided by Polanyi and Szabo since they had dehydrated their pentan-1-ol (n-amyl alcohol) over alumina and then carried out density measurements on the water obtained. Work carried out by G. A. Mills and S. C. Hindin (1950) showed that water exchanges oxygen rapidly with alumina and so the water obtained in the first investigation would have been isotopically normal regardless of the isotopic composition of the alcohol from which it had been obtained. Fortunately, at about this time other workers tested a variety of other reactions so as to obtain evidence for acyl-oxygen fission (wavy line) — rather than alkyl-oxygen fission (dotted line) — postulated by Polanyi some twenty years previously,

$$CH_3-\underset{\underset{O}{\parallel}}{C}+O+C_5H_{11}$$

The next step required chemists to determine the nature of the intermediate and/or transition state formed during the reaction. It was also necessary to establish whether oxygen exchange occurred during hydrolysis rather than without 'reaction'. In basic essentials, the problem required chemists to distinguish between a direct displacement reaction — such as

$$HO^- + \underset{\underset{O}{\parallel}}{\overset{\overset{R}{|}}{C}}-O-R' \rightleftharpoons \left(\overset{\delta^-}{HO} \dots \underset{\underset{O}{\parallel}}{\overset{\overset{R}{|}}{C}} \dots \overset{\delta^-}{O}-R' \right) \rightleftharpoons H-O-\underset{\underset{O}{\parallel}}{\overset{\overset{R}{|}}{C}} + O-R'^-$$

transition state

and a stepwise reaction requiring the formation of an intermediate, as with

$$\underset{\underset{O}{\parallel}}{\overset{\overset{R}{|}}{C}}-O-R' + H-O-H \xrightarrow{OH^-} \left[H-O-\underset{\underset{O-H}{|}}{\overset{\overset{R}{|}}{C}}-O-R' \right] \rightleftharpoons H-O-\underset{\underset{O}{\parallel}}{\overset{\overset{R}{|}}{C}} + H-O-R'$$

intermediate

In 1951, M. L. Bender examined the hydrolysis of a number of esters in water or water–dioxan mixtures and used — in each instance — water enriched in ^{18}O. He compared the rates of exchange of oxygen between the medium and the ester (as determined by the ester recovered after different amounts of partial hydrolysis) with the rate of hydrolysis. He found that the rate of exchange varied from 10% to 40% of the rate of hydrolysis, depending on the ester and the medium used. This result requires the formation of an intermediate which will live long enough to survive the proton shift required to bring about its two non-alkylated oxygen atoms into equivalence so that either might subsequently be split off to regenerate the ester. Of course, in those instances where the alkylated oxygen was split off, the normal products of hydrolysis would be formed. The scheme of reaction which follows summarises this statement of experimental results and interpretation.

$$
\begin{aligned}
&R-\underset{\underset{18}{\overset{\|}{O}}}{C}-O-R' + H-O-H \underset{k_1}{\overset{OH^-}{\rightleftharpoons}} \\
&R-\underset{\underset{O}{\overset{\|}{}}}{C}-O-R' + H-\!\!\overset{18}{}\!\!O-H \underset{k_2}{\overset{OH^-}{\rightleftharpoons}}
\end{aligned}
\left[
\begin{array}{c}
O-H \\
| \\
R-C-O-R' \\
| \\
\overset{18}{}O-H
\end{array}
\right]
\xrightarrow{k_3}
R-\underset{O}{\overset{\|}{C}}-O-H + R'-O-H
$$

intermediate

It was found that the rate constants of concurrent hydrolysis, k_h, and exchange, k_{ex}, did not differ greatly from one another for the systems examined, although in no instance were these rate constants equal. It was also found that the partitioning of the intermediate postulated by the mechanism could be interpreted quite simply

$$k_h/k_{ex} = 2k_3/k_2 \tag{3.6}$$

The results obtained for the exchange and hydrolysis reactions of ethyl benzoate under both alkaline and acid conditions are of particular interest. They are summarised in Table 3. It will be noted that the values of the ratio k_h/k_{ex} differ by a factor of two. However, since the rates of hydrolysis for these two reactions differ by a factor of 10^4 (approximately), there would appear to be some evidence to favour the existence of a common intermediate in these instances (i.e. an intermediate consisting of the unionised hydrate of the ester).

Table 3

Reaction	leaving group	k_h/k_{ex}	k_3/k_2
ethyl benzoate + OH^-	$OC_2H_5{}^-$	11·3	5·7
ethyl benzoate + H_3O^+	C_2H_5OH	5·3	2·7

Other reaction studies have shown that a number of other compounds behave in a similar manner to esters. Concurrent carbonyl oxygen exchange and hydrolysis have been shown to occur not only for simple esters but also for amides, acid anhydrides and acid chlorides. In each case, (3.6) was observed (cf. Table 4 for data).

It is appropriate to conclude this discussion on the hydrolysis of simple esters under alkaline conditions by speculating on the nature of the transition state which leads to the formation of the tetrahedral intermediate shown in the reaction scheme. There would appear to be a number of possibilities. One could argue that

49

$$\text{(I)} \qquad \text{(II)} \qquad \text{(III)}$$

(I) might be an appropriate formulation since it resembles the intermediate to which it leads extremely closely. However, the composition of the transition state could be more complex and may involve water. Thus, (II) requires the formation of a cyclic complex with one molecule of water in addition to the OH^- ion whereas (III) requires the use of two molecules of water in addition to the OH^- ion to form a similar cyclic complex. As yet, it has not been possible to distinguish between these possibilities, (I), (II) and (III).

Table 4

Reaction in aq. soln.	leaving group	k_h/k_{ex}	k_3/k_2
benzamide + OH^-	NH_2^-	0·53	0·27
benzamide + H_3O^+	NH_3	no exchange reaction	
benzoic anhydride + H_2O (using 75% dioxane-water)	$O_2C.C_6H_5^-$	20	10
benzoyl chloride + H_2O (using 75% dioxane-water)	Cl^-	25	12·5

Decomposition of hydrogen peroxide

Hydrogen peroxide solution decomposes extremely slowly in the absence of a catalyst. A number of solids, such as platinum black or manganese(IV) oxide, and a number of ions, such as iron(II) or bromide ions in the presence of aqueous acid, catalyse the reaction. The stoichiometric equation for the oxidation of bromide ions by hydrogen peroxide in dilute aqueous acid solution is

$$2Br^-(aq) + H_2O_2 + 2H^+(aq) = Br_2 + 2H_2O$$

The rate equation for this reaction is given by

$$v = k[H_2O_2][H^+][Br^-] \tag{3.7}$$

The mechanism for the reaction can be interpreted in a number of ways. The simplest requires the use of *at least* two steps:

$$(1) \quad H^+ + Br^- + H_2O_2 \rightarrow HOBr + H_2O \text{ (slow)}$$

(2) $HOBr + H^+ + Br^- \rightleftharpoons Br_2 + H_2O$ (fast)

The sum of (1) and (2) yields the stoichiometric equation and the known properties of the postulated intermediate, HOBr, conform to the expected nature of an intermediate, namely of a substance which is not very stable. Indeed if a mixture of hypobromous acid (bromic(I) acid) and bromide ions in water is acidified, then bromine is produced quite rapidly. From (3.7), it follows that the composition of the transition state is (H_2O_2, H^+, Br^-) and a possible structure for this species is

$$\begin{pmatrix} H & & & Br \\ & \diagdown O \cdots O \diagup & \\ & \diagup & & \diagdown \\ H & & & H \end{pmatrix}$$

A similar account of the decomposition of hydrogen peroxide by iodide ions in the presence of dilute aqueous acid can be devised, (cf. Experiment on p. 22).

EXPERIMENT   (1) Investigate the decomposition of hydrogen peroxide by a mixture of iron(II) and iron(III) ions.

(2) Make a comparative study of the catalytic action of the ions of transition metals on the decomposition of hydrogen peroxide.

(3) Examine the decomposition of hydrogen peroxide by manganese(IV) oxide with the view to determining the mechanism of this reaction.

(4) Investigate the reaction between hydrogen peroxide and halide ions in the absence of hydrogen ions (cf. p. 22).

Fast reactions

The rates of many chemical reactions have been measured by what must now be considered to be conventional techniques. As we have seen, one could start a reaction by mixing, for example, two solutions and then measure the concentrations of one or more of the species present at known intervals of time. Unfortunately, this type of procedure cannot be used if the start of a reaction is ill defined when the time of mixing is compatible with or is greater than the total time required for the reaction to occur. A similar constraint will apply if the time required to complete a measurement on a reaction mixture is no longer negligible when compared with the duration of the reaction.

The first difficulty can sometimes be overcome by using a *flow system*. For example, if one wanted to follow the reaction between species A and B in solution, samples of the two solutions could be led to a mixing chamber before being passed through an observation tube. The concentration of one or more of the species present could then be detected at various points along the observation tube by spectroscopic means and recorded by some form of high speed recorder. Chemical events with half-lives down to about 10^{-3} s have been studied by this means.

Yet other chemical systems, such as the neutralisation of hydrogen ions by hydroxide ions, require an even shorter period of time before reaction is complete and have been investigated by so called *relaxation methods*. These procedures start with a chemical system in 'equilibrium'. The system is then disturbed by using an ultrasonic wave or by using an alternating electric field and the course of reaction is then followed as it again approaches a state of equilibrium, again using high speed recording techniques. For example, the ionisation of acetic acid (ethanoic acid) may be written as

$$CH_3.CO.OH + H_2O \; \underset{k_2}{\overset{k_1}{\rightleftharpoons}} \; CH_3.CO.O^- + H_3O^+$$

and the values of the rate constants k_1 and k_2 determined from the results of experiments using relaxation methods. At $25°C$, k_1 has the value $7.8 \times 10^5 \text{ s}^{-1}$ and k_2 the value $4.5 \times 10^{10} \text{ dm}^3 \text{ mol}^{-1} \text{ s}^{-1}$.

Shock tube methods have been applied to the study of fast reactions, especially to those in the gas phase. Essentially, an inert gas at high pressure in a tube is separated from a reactive gas at low pressure by a diaphragm. When this diaphragm is broken by some mechanical means, the expansion of the inert gas produces a shock wave which passes through the low pressure gas compressing it adiabatically. The result is the production of an extremely rapid change in temperature and the subsequent reaction can often be followed by spectroscopic means.

Flash photolysis is another means of producing an extreme disturbance in a pre-mixed chemical system and is of relevance to reactions activated by light. Intense light energy (in an appropriate region of the spectrum) is released for an extremely brief moment of time $(c \cdot 10^{-6} \text{ s})$ and causes both electronic excitation and chemical reaction. Usually the kinetics of excited molecules or free radicals in the period immediately following the flash are observed by means of subsequent light absorption or emission using photographic or photoelectric techniques. The method has been used to study reactions in the gas phase and in solution.

To avoid the difficulty of interpreting reactions between molecules which possess a variety of energies dependent upon a distribution law, it is now feasible to project molecules with approximately the same velocity as a *molecular beam*. By using two such molecular beams and arranging for them to interact, a chemical reaction can be made to occur. In this way the kinetics of certain bimolecular reactions can be studied in fine detail. The technique has led to significant conclusions about the manner in which energy is distributed between molecular species during a reaction.

4 Some theoretical interpretations of reaction rates

Simple collision theory and reactions in the gas phase

It is appropriate to seek some theoretical basis for the qualitative ideas expressed in earlier parts of this book. Ideas such as those expressed by Fig. 1 need to receive consideration before one attempts to account for more complex situations, such as those described in Chapter 3.

Any attempt to interpret gas phase reactions must clearly depend on the kinetic theory of gases and on such general considerations of the energetics of reaction as may be taken to be self-evident. The rate of the forward reaction must depend in some way on the energy possessed by the particles taking part in the process. In the case of a reaction in the gas phase, we might expect to find a wide distribution of molecular speeds and therefore of energies. Even if we could start off with all of the molecules of a gas travelling in different directions with equal speeds, it would not be long before random collisions occurred between the molecules thereby increasing the speed of some and retarding others. We might expect to find the speeds of the gas molecules ranging from zero to an extremely high value but grouped around some average value determined in some way by the temperature. This theoretical problem was solved more than a century ago by J. C. Maxwell at King's College London and by L. Boltzmann in Vienna, working independently. This was long before any experimental verification was feasible.

Since then, various experimental procedures have been used to determine the distribution of speeds in a gas. All require special apparatus and facilities which are not usually available in a laboratory. Perhaps the experiment devised by I. F. Zartman and C. C. Ko (1931–34) is the simplest to understand. The experimental difficulties were great: it was necessary to operate the apparatus in high vacuum, to produce particles of the same mass and to release them into the vacuum in such a manner as not to alter the natural distribution of speed between the particles so released. A diagrammatic representation of Zartman and Ko's apparatus is shown in Fig. 25. It consists of a hollow cylinder possessing a slit parallel to its axis and rotating in a vacuum at high speed. An oven containing bismuth was used as a source of particles and a beam of such particles was collimated by a series of slits. The collimated beam of particles became aligned with the slit in the cylinder once every revolution. Under these conditions, particles entered the cylinder and were deposited on a glass plate. Due to differences in speed, these particles were found not to strike the plate in the same place but were spread out into a band. Measurements of the intensity of this band by means of a photometer gave an indication of the distribution of speeds of the particles which entered the cylinder. At 800°C, it was found that the band of bismuth particles was spread out over some 3 cm on the glass plate and the range of speeds for the particles was calculated to be from 168 m s^{-1} to 673 m s^{-1}. These results were then shown to follow a definite pattern, the Maxwell–Boltzmann distribution (cf. Fig. 26), within the limits imposed by the experiment. Other experimental determinations confirmed this

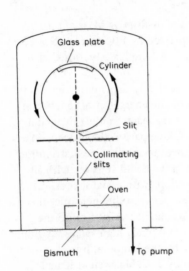

Fig. 25

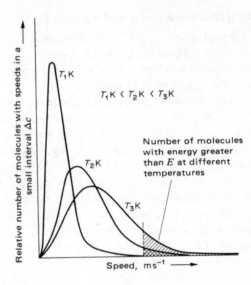

Fig. 26

result whereas others have demonstrated the manner in which the distribution varies with temperature.

The reader is invited to show that, using kinetic theory, the *average* translational kinetic energy of one mole of gas is

$$\bar{E}_{trans} = 3\,RT/2 \qquad (4.1)$$

where R is the gas constant and T the temperature on the Kelvin scale. It will be noted that this average energy is independent of the mass of the molecule. For a monatomic gas — such as helium, argon or mercury vapour — the translational kinetic energy is essentially the only energy possessed by the gas particles. For diatomic gases — such as oxygen or nitrogen — and for polyatomic gases — such as methane, ammonia or sulphur dioxide — energy associated with other motions within the molecule has to be taken into account. Rotation and vibration store energy to an extent dependent upon the number of bonds and atoms in a molecular species. Accordingly, the *average* effective energy of one mole of polyatomic molecules is

$$\bar{E} = \bar{E}_{trans} + \bar{E}_{rot} + \bar{E}_{vib} \qquad (4.2)$$

where $\bar{E}_{trans}$ refers to average translational energy, $\bar{E}_{rot}$ refers to average rotational energy and $\bar{E}_{vib}$ refers to average vibrational energy. Of course, all molecules also possess electronic energy but at ordinary temperatures the amount each molecule possesses is almost independent of temperature and, for our present purpose, can be neglected. Perhaps the really significant fact which follows from the Maxwell-Boltzmann distribution when applied to translational energy is that the fraction of molecules x_E which possesses energies greater than a value E at temperature T approximates closely to $\exp(-E/RT)$.

$$x_E = e^{-E/RT} \qquad (4.3)$$

Fig. 26 shows that the fraction x_E for a given temperature T is quite small. This value will increase with temperature and the actual number of collisions between molecules possessing an energy greater than the limiting value, E, must also increase. Now if we assume that the colliding molecules need to possess energy equal to or in excess of a critical value, then the concentration of such molecules in the reaction mixture should determine the rate of reaction, i.e. that the rate of reaction depends not only on the number of collisions per second in a unit volume but also on the chance that colliding pairs of molecules possess an energy of E or more. (This hypothesis is not the only possible hypothesis: we might require simply that reaction probability is some function of energy. However, the first suggestion provides an approach to quantitative description.) Accordingly, we could state that the rate of reaction, v, is given by

$$v = \text{(collision rate)} \times e^{-E/RT}. \qquad (4.4)$$

In Chapter 2, a general rate equation for a second order reaction was derived and the principle applied to a number of reactions. In Chapter 3, it was applied to a simple bimolecular process, the decomposition of hydrogen iodide (cf. p. 32) where it was found that

$$v = k_2 \, [\text{HI}]^2$$

for the reaction

$$2\text{HI} \rightarrow \text{products}$$

We can now modify (4.4) and relate it to an expression of the rate of reaction using concentration terms. For example, a simple reaction between two species A and B might be

$$\text{A} + \text{B} \rightarrow \text{products}$$

and the rate of this reaction is proportional to the product of the concentrations of the two reactants

$$v = k_2 [A] [B]$$

where k_2 is a second order rate constant. It follows that

$$k_2 [A] [B] = (\text{collision rate}) \times e^{-E/RT}, \tag{4.5}$$

and that the rate constant, k_2, is the rate when the reactants are at unit concentrations

$$k_2 = \left(\begin{array}{c} \text{collision rate when} \\ \text{reactants are at unit} \\ \text{concentrations} \end{array} \right) \times e^{-E/RT}. \tag{4.6}$$

Equation (4.6) is of the same form as the Arrhenius equation (2.24) and equates the constant term A of that equation with the frequency of collisions when the reactants are at unit concentrations.

Some support for the above argument can be obtained by applying the ideas used to an actual situation. For example, the decomposition of hydrogen iodide yields a rate constant $k_2 = 7 \cdot 04 \times 10^{-7}$ dm^3 mol^{-1} s^{-1} at 556 K. If we assume that the concentration of hydrogen iodide is $0 \cdot 01$ mol dm^{-3}, we can calculate the number of effective collisions which take place each second at 556 K. The rate of reaction is

$$-d[HI]/dt = k_2 [HI]^2$$
$$= 7 \cdot 04 \times 10^{-7} \times [10^{-2}]^2.$$

Since the number of molecules present in one mole is given by $L = 6 \cdot 02 \times 10^{23}$ mol^{-1}, it follows that the number of molecules of hydrogen iodide decomposing in each second at 556 K is $7 \cdot 04 \times 10^{-11} \times 6 \cdot 02 \times 10^{23}$. Accordingly, the number of *effective* collisions must be $\frac{1}{2} \times 7 \cdot 04 \times 10^{-11} \times 6 \cdot 02 \times 10^{23} = 2 \cdot 1 \times 10^{13}$. However, the *actual* number of collisions per second can be obtained from kinetic theory since the average speed of the molecules of a gas can be calculated from

$$\bar{c} = (8RT/\pi M)^{\frac{1}{2}}$$

where M is the molecular mass of the gas and R is the gas constant. Thus, at 556 K, the average speed of the molecules of hydrogen iodide is 300 m s^{-1}. Now under these conditions the mean free path of the hydrogen iodide molecule is about 3×10^{-7} m and so in 10^{-2} mol of hydrogen iodide in 1 dm^3 there must be

$$\frac{3 \cdot 0 \times 10^2 \times 10^{-2} \times 6 \cdot 02 \times 10^{23}}{3 \times 10^{-7} \times 2} = 3 \cdot 01 \times 10^{30} \text{ collisions s}^{-1}.$$

Thus, it follows that only 1 in $1 \cdot 5 \times 10^{17}$ collisions is an *effective* collision under the conditions used.

We can take the matter much further if the rate constants (at known temperatures) and the energy of activation are known. Thus for the decomposition of hydrogen iodide, the rate constants at two specific temperatures are $k_{2(556\ K)} = 7.04 \times 10^{-7}$ dm^3 mol^{-1} s^{-1} and $k_{2(666\ K)} = 4.38 \times 10^{-4}$ dm^3 mol^{-1} s^{-1} and the energy of activation is 184 kJ mol^{-1}. Accordingly, using (4.3)

at 556 K $\qquad x_E = \exp(-184\ 000/8.314 \times 556)$

$\qquad\qquad\qquad = 5.2 \times 10^{-18};$

at 666 K $\qquad x_E = \exp(-184\ 000/8.314 \times 666)$

$\qquad\qquad\qquad = 3.7 \times 10^{-15}.$

So, the ratio of the two rates will be given by

$$\frac{\text{Rate of reaction at 666 K}}{\text{Rate of reaction at 556 K}} = \frac{x_E \text{ at 666 K}}{x_E \text{ at 556 K}} = 7.1 \times 10^2.$$

The corresponding ratio of the two rate constants yields a similar (and almost equivalent) result:

$$\frac{k_2 \text{ at 666 K}}{k_2 \text{ at 556 K}} = 6.2 \times 10^2,$$

which lends some support for the approximations adopted.

The above account of reaction rate expressed in terms of a simple collision mechanism follows directly from the concept of reaction mechanism shown in Fig. 1 and is supported by the detailed application of theory to simple bimolecular reactions, (as with the decomposition of hydrogen iodide). This type of reaction permits some simplification to be made and other considerations will be necessary before one can accept the theory as possessing wide application. A variety of different types of reaction will need to be considered in detail, including different types of gas phase reaction and reactions in solution. In addition, some consideration as to what actually occurs during an effective collision ought to be included in the survey.

Reactions in solution

A simple consideration of reactions in solution can be achieved by assuming that the solvent molecules act as a physical barrier to the diffusion of reactants. A comparison of this model of a reaction in solution with the same reaction in the gas phase is possible from Fig. 27. Solvent particles have been marked X and reactants are assumed to be spherical particles and are marked A or B. All three species are assumed to be comparable in size. Given that the concentrations of A and B are the same in both instances and there is sufficient free space for the particles in solution to diffuse, it can be shown that there is very little difference between the number of collisions per second in these two cases. In the instance provided by the left hand

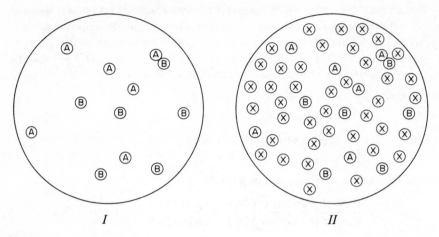

Fig. 27. A comparison of a reaction taking part in (I) the gas phase and (II) solution, using the same concentrations of reactants A and B

portion of Fig. 27, each particle A will have single collisions with very many different particles. In the other example, each particle A will experience a number of repeated collisions with a smaller number of particles. The factors responsible for decreasing the ease of diffusion of reactant particles A and B towards one another are also responsible for decreasing the ease of diffusion of these particles (and those produced by reaction) away from one another. Accordingly, once two reactant particles A and B have met, they will most likely remain in relatively close proximity until sufficient energy has been acquired for reaction to occur. The 'cage' effect of solvent molecules must also apply to the products of reaction. This simple model of chemical reaction leads one to expect reactions in solution to be somewhat slower than those in the gas phase.

Simple collision theory has been applied to a number of relatively slow reactions in solution. For example, the solvolysis of bromoethane under alkaline conditions in ethanol yields a calculated rate constant which is 0·9 of that obtained by experiment. Such good agreement with theoretical calculations is not always obtained. Indeed, reactions which are diffusion controlled (as with precipitation reactions like the formation of barium sulphate) do not support the approach which has been outlined in this section. (Theoretical treatments of diffusion controlled reactions are beyond the scope of this introduction to kinetics. Some additional information about these reactions appears in the more advanced texts listed in the bibliography.)

Unimolecular reactions
A number of gas phase reactions observe first order kinetics and our discussion so far has ignored this type of reaction. It would appear reasonable to suggest that any energy of activation required for such reactions must come from kinetic energy

transferred during collisions. However, as will be appreciated, the rate of reaction cannot depend on collision frequency in these cases since the rate is proportional to the first and not the second power of the concentration of reactant. J. A. Christiansen and F. A. Lindemann (1921), working independently, showed how collision mechanisms could lead to first order kinetics. As has already been noted (cf. p. 21 and p. 38), systems which exhibit such behaviour include relatively complex molecules.

When a molecule absorbs energy, it can execute a large number of different types of vibration and many of these will be unimportant as far as chemical transformation is concerned. However, if such a molecule acquired energy as the result of collision processes, a chemical change might take place if the critical quantity of energy required for the process were so organised within the molecule as to lead to the rupture of a bond. Accordingly, it would appear that for a reaction of this type, atomic rearrangements take place as a consequence of collisions rather than during collisions. It follows that a suitably energised molecule could move away from the site of collision before any energy re-organisation could occur and that the proportion of such molecules which actually undergo decomposition could be quite small.

We can picture the situation by representing normal molecules by the symbol A and 'activated' molecules by A*. Normal molecules take part in a collision and can produce an activated molecule which can then decompose to form the products of reaction.

$$A + A \underset{k_{-2}}{\overset{k_2}{\rightleftharpoons}} A + A^* \\ \downarrow k_1 \\ B + C$$

The rate of formation of A* will be

$$d[A^*]/dt = k_2 [A]^2 - k_1 [A][A^*] - k_{-2}[A][A^*]. \qquad (4.7)$$

The rate of disappearance of A will be given by

$$-d[A]/dt = k_2 [A]^2 - k_{-2}[A][A^*]. \qquad (4.8)$$

The rate of formation of product B will be given by

$$d[B]/dt = k_1 [A^*]. \qquad (4.9)$$

When the reaction has been going for a short time, the rate of formation of activated molecules, A*, will equal the rate of their disappearance. This approximation is known as the steady state approximation and simplifies the mathematics required to relate equations (4.7), (4.8) and (4.9).

It follows that when

$$d[A*]/dt = 0,$$

equation (4.7) becomes

$$[A*] = k_2 [A]^2 / (k_{-2} [A] + k_1)$$

and so, using (4.9),

$$d[B]/dt = k_1 k_2 [A]^2 / (k_{-2} [A] + k_1) \qquad (4.10)$$

Using (4.10), two situations can be devised to provide tests for this 'model' of unimolecular reactions:

(a) the situation where the rate of decomposition of A* is considered to be much greater than its rate of deactivation: i.e. when $k_1 \gg k_{-2} [A]$. Under these conditions, the rate of formation of product B will be given by

$$d[B]/dt = k_2 [A]^2. \qquad (4.11)$$

(b) the situation where the rate of deactivation of A* is considered to be much greater than its rate of decomposition: i.e. when $k_{-2} [A] \gg k_1$. Under these conditions, the rate of formation of product B will be given by

$$d[B]/dt = k_1 [A] \qquad (4.12)$$

These special situations can be tested experimentally. Thus, if the pressure of a gaseous system were decreased, the rates of both activation and of deactivation must also decrease. At a low enough pressure, the condition for first order kinetics must always fail when $k_{-2} [A]$ is no longer very much greater than k_1. Accordingly one might expect a first order rate constant to decrease at low pressures and eventually change to a second order rate constant. Experiments on the isomerisation of cyclopropane at various pressures have been found to yield such data.

Clearly if all of this behaviour is the result of a lowered probability of activation and deactivation, it should be possible to restore the original state of affairs by merely adding sufficient pressure of a completely inert gas. This has been attempted and the results obtained agree with the theoretical forecast. Thus, the isomerisation of cyclopropane has provided much useful data through the addition of gases such as helium, argon, water vapour and propylene (propene).

Another theoretical problem

The interpretation of the rate of reaction in terms of a collision process can be checked by calculations and many systems provide Arrhenius A-factors which, in principle, could be calculated. Indeed this has been attempted and quite good agreement between theoretical and experimental values are obtained for systems such as the decomposition of hydrogen iodide (cf. p. 57). Unfortunately, such agreement is not universal for all systems and variations occur over a wide range.

To account for such variations, modifications to the theory have been tried, including the use of a factor to permit due allowance to be made for the correct positioning of molecules in a collision. In essence, this modification suggests that the Arrhenius A-factor is a product of two terms, the collision frequency, Z, and the steric factor, P. Hence, (2.24) becomes

$$k = PZ\, e^{-E/RT} \tag{4.13}$$

If one reviews the large number of chemical systems where this has been used, values given to this P-factor vary over a very wide range indeed. The P-factor appears to be little more than a simple empirical correction in many instances rather than a function of the geometry of the system undergoing a collision process. To account for variations in the PZ term of (4.13), we need to adopt another approach to the elementary processes under review.

An approach to transition state theory

Theoretical ideas stemming from a simple collision theory of reaction rates can account for a large number of effects found during a study of chemical reactions. However, no account has been given of the nature of an 'effective' collision. Secondly, the interpretation of a reaction in the manner shown in Fig. 6 finds wide application where the factor controlling the rate of reaction in all instances depends upon the occurrence of sufficient acts of activation. Indeed, we have assumed that the transition state exists only as the result of an 'effective' collision and that as soon as vibrations take place in a particular mode, the complex decomposes. The reaction between two species A and B, passing through a transition state, TS, to form products, may be written as

$$A + B \rightleftharpoons TS \rightarrow products$$

If we assume that the transition state complex TS is in equilibrium with the reactants, then the concentration of TS will be given by

$$[TS] = K^{\ddagger}\,[A]\,[B] \tag{4.14}$$

where $K^{\ddagger}$ is the equilibrium constant for the formation of the transition state complex, TS.

To be able to forecast the behaviour of the transition state as a species, we now need to assume that it has many — if not most — of the properties of a normal chemical species. It will, for example, decompose when sufficient energy exists in at least one of its bonds. This must occur when the vibrational energy of that bond is such that the amplitude of vibration is 'excessive'. The rate of reaction may be expressed in the following way

$$\text{Rate} = -d[A]/dt = [TS] \times (\text{rate of passage over energy barrier})$$
$$= [TS] \times \nu \tag{4.15}$$

61

where ν is the frequency at which the 'weak' bond in the transition state complex will rupture. Equation (4.15) can lead to a more convenient form of rate equation (4.19) using the arguments and ideas of statistical mechanics. However, for our present purpose, a simpler but less rigorous approach may be acceptable.

Equation (4.2) states a general truth relating to the partition of energy in a polyatomic molecule and kinetic theory shows that the average translational energy for one molecule is

$$\bar{E}_{trans} = 3kT/2 \tag{4.16}$$

where k is the Boltzmann constant. Since there are three components to this translational energy, it would not be unreasonable to assume an equipartition of energy between these components such that in any one there is $\frac{1}{2}kT$. The *principle of equipartition of energy* may be stated as an extension of this idea: the total energy possessed by a molecule (or other species) is divided equally between the three energies of translation, the various energies of rotation, and, if the temperature is sufficiently high, the two components of each energy of vibration. (One may conveniently consider a vibration to consist essentially of energy with a potential and a kinetic component.) It follows that, under these conditions, the average energy of vibration will be kT.

M. Planck (1900) put forward the view that the energy of vibration could be expressed in the form

$$E = h\nu \tag{4.17}$$

where ν is the frequency of vibration and h is a constant, now referred to as Planck's constant. If we adopt (4.17) for the transition state complex, it follows

$$h\nu = kT \tag{4.18}$$

and this result can be used to modify (4.15) as follows

$$\text{rate} = -d[A]/dt = [TS] \times (kT/h). \tag{4.19}$$

Using (4.14), this becomes

$$-d[A]/dt = (kT/h)K^{\ddagger}[A][B] \tag{4.20}$$

which is of the same form as rate equations used in earlier parts of this book – e.g.
$$-d[A]dt = k_2[A][B]$$

where k_2 is a constant, the second order rate constant for the reaction between reactants A and B.

PROBLEM   Use (4.18) to determine the value of ν over a series of temperatures. ($h = 6 \cdot 6256 \times 10^{-34}$ J s; $k = 1 \cdot 38054 \times 10^{-23}$ J K^{-1}). Are the values obtained consistent with those which one might expect for 'weak' chemical bonds?

The argument which has been put forward depends on a number of assumptions, including that of an equilibrium between the reactants and the transition state. It is of interest to speculate whether one can interpret this pseudo-equilibrium using pseudo-thermodynamic terms and so obtain from (4.20) an interpretation of the Arrhenius parameters A and E. By adapting relationships which apply to chemical equilibria, we can write

$$K^{\ddagger} = \exp(-\Delta G^{\ddagger}/RT)$$

and

$$\Delta G^{\ddagger} = \Delta H^{\ddagger} - T\Delta S^{\ddagger},$$

where $\Delta G^{\ddagger}$ is the free energy of activation, $\Delta H^{\ddagger}$ the enthalpy of activation, and $\Delta S^{\ddagger}$ the entropy of activation. These relationships can be applied to (4.20):

$$-d[A]/dt = (kT/h)[A][B]\exp(-\Delta G^{\ddagger}/RT)$$
$$= \{(kT/h)\exp(\Delta S^{\ddagger}/R)\}[A][B]\exp(-\Delta H^{\ddagger}/RT) \quad (4.21)$$

which has the form of an earlier rate equation

$$-d[A]/dt = A[A][B]\exp(-E/RT) \quad (4.22)$$

and embraces the Arrhenius equation (2.24) in place of the rate constant.

It will be appreciated that (4.21) does not require collision frequencies or steric factors to be estimated when one calculates the Arrhenius parameter A, (cf. (4.13)). However, a simple comparison of (4.21) and (4.22) shows that it is still necessary to estimate one quantity, $\Delta S^{\ddagger}$:

$$A = (kT/h)\exp(\Delta S^{\ddagger}/R) \quad (4.23)$$

Transition state theory gives the equation

$$k = (kT/h)\exp(\Delta S^{\ddagger}/R)\exp(-\Delta H^{\ddagger}/RT) \quad (4.24)$$

where k is the rate constant. Equation (4.24) can be written in the form

$$\ln k = \ln(kT/h) + \ln T + \Delta S^{\ddagger}/R - \Delta H^{\ddagger}/RT.$$

On differentiating,

$$d(\ln k)/dT = 1/T + \Delta H^{\ddagger}/RT^2. \quad (4.25)$$

The differential form of the Arrhenius equation (2.24) is

$$d(\ln k)/dT = E/RT^2, \quad (4.26)$$

hence,

$$E = \Delta H^{\ddagger} + RT.$$

It is apparent that there is relatively little difference between the Arrhenius energy of activation, E, and the enthalpy of activation, $\Delta H^{\ddagger}$, when one compares the order of magnitude of typical energies of activation with values for the term RT.

On using (4.27) and (4.24), we obtain

$$k = (kT/h) \exp(\Delta S^{\ddagger}/R) \exp(-\Delta H^{\ddagger}/RT)$$

$$\approx (kT/h) \exp(\Delta S^{\ddagger}/R) \exp(-E/RT).$$

Taking logarithms,

$$2 \cdot 303 \lg k = 2 \cdot 303 \lg(k/h) + 2 \cdot 303 \lg T + (\Delta S^{\ddagger}/R) - (E/RT)$$

Rearranging,

$$\Delta S^{\ddagger} = 2 \cdot 303 R(\lg k - \lg(kT/h) - \lg T) + E/T \qquad (4.28)$$

which enables $\Delta S^{\ddagger}$ to be estimated if the rate constant k and the energy of activation E are known. (It is also worth noting that the same experimental data can be used to determine the Arrhenius A-factor.)

For a reaction between simple molecules or between atoms, there will be but a small re-organisation of energy in the species known as the transition state complex when compared with that in the original reactants. Accordingly, we might expect $\Delta S^{\ddagger}$ to possess a small positive or negative value. For reactions between more complex molecules, considerable reorganisation may be necessary and so one might expect larger numerical values for the entropy of activation. A 'positive' value for $\Delta S^{\ddagger}$ will indicate a 'normal' reaction: a 'negative' value, a reaction which is relatively 'slow'. Calculations for actual systems tend to bear out these general expectations.

A comparison of theoretical treatments of reaction rates

In principle, both collision theory and transition state theory are correct and, when applied correctly, they can yield consistent results. Any attempt to evaluate their relative merits must therefore be based on a comparison of their usefulness in analysing practical problems, remembering that the purpose of a theory is to correlate experimental data in a simple manner and to predict the results of other experiments which have yet to be carried out. Both theories have achieved notable success in correlating data from a wide range of experiments. Perhaps the chief distinction between the two theories stems from the change in emphasis given to the collision process. In collision theory, the collision process is the key feature whereas in transition state theory the events which follow from a collision form the key feature. In recent years, the majority of successful predictions regarding reaction rates have been achieved through the use of transition state theory. The essentially qualitative description of a simple reaction given in Chapter 1 has an essentially semi-quantitative basis for certain elementary reaction processes. It has also formed an important turning point in the development of the theory of reaction rates.

Bibliography

Introductory reading

COE, J. S. *Chemical Equilibrium: An Introduction.* Methuen, 1971
DAWSON, B. E. *Energy in Chemistry: An Approach to Thermodynamics.* Methuen, 1971
ROBERTSON, A. J. B. *Catalysis in Chemistry.* Methuen, 1972

Further reading

ASHMORE, P. G. *Principles of Chemical Equilibrium.* Royal Institute of Chemistry, Monographs for Teachers, No. 5, 1961
ASHMORE, P. G. *Principles of Reaction Kinetics.* Royal Institute of Chemistry, Monographs for Teachers, No. 9, 1965
BOND, G. C. *Principles of Catalysis.* Royal Institute of Chemistry, Monographs for Teachers, No. 7, 1963 and 1968
STEWART, R. *The Investigation of Organic Reactions.* Prentice-Hall, 1966
SYKES, P. *A Guidebook to Mechanism in Organic Chemistry.* Longmans, 1961 or subsequent editions
WYATT, P. A. H. *The Molecular Basis of Entropy and Chemical Equilibrium.* Royal Institute of Chemistry, Monographs for Teachers, No. 19, 1971

Reference Books

A number of textbooks include a more advanced treatment of parts of the work described in this book, including the following:

DEWAR, M. J. S. *An Introduction to Modern Chemistry.* Athlone Press, University of London, 1965
FROST, A. A. and PEARSON, R. G. *Kinetics and Mechanism: A study of Homogeneous Chemical Reactions.* John Wiley and Sons, 1953 and 1961
HAGUE, D. N. *Fast Reactions.* Wiley-Interscience, 1971

A number of general textbooks also provide additional information:

BARROW, G. M. *Physical Chemistry.* McGraw-Hill Book Co., 1961
GLASSTONE, S. *Textbook of Physical Chemistry.* 2nd edition. Van Nostrand, 1946
MOORE, W. J. *Physical Chemistry.* 5th edition. Longmans, 1972

HAMMETT, L. P. *Physical Organic Chemistry.* 2nd edition. McGraw-Hill Book Co., 1970

INGOLD, C. K. *Structure and Mechanism in Organic Chemistry.* Bell, 1953 and 1969

LOCKHART, J. C. *Introduction to Inorganic Reaction Mechanisms.* Butterworths, 1966

NORMAN, R. O. C. *Principles of Organic Synthesis.* Methuen, 1968

NORMAN, R. O. C. (ed) *Essays on Free Radical Chemistry.* Special Publication No. 24. The Chemistry Society, 1970

RIDD, J. H. (ed) *Studies on Chemical Structure and Reactivity.* Methuen, 1966

Index

Heterogeneous catalysis, 29
Heterolytic fission, 37
Hindin, S. C., 48
Homogeneous catalysis, 29
Homolytic fission, 37
Hughes, E. D., 41, 43, 45
Hughes-Ingold terminology, 45
Hydrogen atom reactions, 3, 5, 6, 9
Hydrogen iodide, decomposition, 26, 28, 32, 55, 57, 60
Hydrogen-iodine reaction, 32
Hydrogen molecule decomposition, 4
Hydrogen peroxide, decomposition, 14, 22, 28, 50
Hydrolysis, of esters, 12, 15, 47
    of halides, 14, 39

Ingold, C. K., 42, 45
Integration of rate equations, 17, 18
Intermediate, reaction, 30, 31, 43, 45, 48, 49, 50, 51
Iodide-hydrogen peroxide reaction, 14, 22, 28, 51
Iodination of acetone, 14, 19, 29
2-Iodo-2-methylpropane, hydrolysis, 21
2-Iodooctane, racemisation, 46
Ion-pair formation, 43, 45
Isotopic tracer technique, 11, 46, 47, 49

Ko, C. C., 53

Lindemann, F. A., 59

Maccoll, A. 37, 38
Maxwell, J. C., 53
Maxwell distribution law, 54
Methyl acetate, hydrolysis, 47
Methyl formate, hydrolysis, 12, 14, 47
Mills, G. A., 48
Models, scientific, 1, 27, 28
Molecular beam, 52

Olefine, see alkene
Order of reaction, 15, 17, 18, 21, 32, 36, 40, 56, 60

P-factor, 61
Pentan-1-ol, dehydration, 48

n-Pentyl acetate, hydrolysis, 47
Persulphate-iodide reaction, 20
Planck, M., 62
Polanyi, M., 47
Potential energy, curve, 4, 5, 7
    surface, 7
Problems, 22, 25, 27, 62

Rate constant, 17, 18, 41, 49, 56, 57
Reaction, exchange, 5
    gas phase, 2, 20, 36, 38, 53
    half-life, 21
    in solution, 57
    mechanism, 2, 6, 10, 12, 21, 30, 45, 57
    model, 3, 9
    pathway, 2, 7, 8, 9, 10, 28, 31
Reaction rate, factors affecting, 3, 12, 15, 25, 28, 55, 61
    measurement, 12
    solvent effects, 42, 57
    temperature and, 25
    theoretical treatments, 13, 53, 64
Relaxation methods, 52

Second order reactions, 17, 18, 22, 40
Shock tube method, 52
S_N1 reaction, 40, 41, 47
S_N2 reaction, 40, 46, 47
Srinivasan, R., 33
Steady state approximation, 59
Steric factor, 3, 61
Sullivan, J. H., 33, 34
Szabo, A. L., 47

Termolecular process, 21
Transition state, 7, 8, 9, 30, 31, 35, 36, 37, 43, 46, 48, 49, 51, 61
Transition state theory, 6, 7, 61

Unimolecular reaction, 41, 43, 58
Urey, H. C., 48

Winstein, S., 44

Young, W. G., 44

Zartman, I. F., 53
Zero order of reaction, 19